Disaster Education and Management

Disaster Education and Management

Editor

Manzar Khan

scitus
academics

Disaster Education and Management

Edited by **Manzar Khan**

Printed in 2017

ISBN: 978-1-68117-125-8

Library of Congress Control Number: 2015936547

SCITUS Academics LLC,
616, Corporate Way, Suite 2, 4766,
Valley Cottage, NY 10989

www.scitusacademics.com

Contents

Preface

Disaster management is the discipline dealing with the organization and management of resources and responsibilities to cater with all humanitarian aspects of emergencies and disasters- both natural and man-made. It involves preparedness, response and recovery in order to lessen and mitigate the impact of disasters. Prevention is better than cure, is a novel saying which is very apt in the context of disaster management. It is now becoming increasingly evident and mitigation and investment in disaster preparedness can save thousands of lives, vital economic assets, livelihoods and reduce the cost of overall disaster relief. There is a never ending role of education and schools in disaster risk reduction. The importance of education in promoting and enabling Disaster Risk Reduction (DRR) has already been identified by researchers and policy makers. In doing so, there is a renewed focus on disaster risk education in primary and secondary schools.

Editor

Chapter 1

Human Response to Natural Disasters

Dara Nix-Stevenson[1]

[1]The University of North Carolina at Greensboro, USA

ABSTRACT

This study elaborates on the connection between socioeconomic status, education, and the ability to respond to natural disasters. Using the aftermath of Hurricane Katrina and other natural disasters as teachable moments, I foreground how uneven access to resources and capital leave some people more vulnerable than others to natural disasters and how marginal communities inevitably bear the accompanying repercussions of who gets what, when, and how much in the postdisaster emergency relief and reconstruction phase. This occurs not necessarily and merely through a "natural" disaster, as the Boxer Day Tsunami or Hurricane Katrina, but through processes of social, political, and economic disempowerment associated with prior racialized histories and inequitable access to cultural capital.

Ecological disasters like the U.S. naval occupation of Vieques, Puerto Rico, 1941-2003; the Indian Ocean Tsunami in December 2004; and Hurricane Katrina in August 2005 both underscore and interrogate how vulnerability is conceptualized in Western discourse (Bankoff, 2001). "Vulnerability, in the disaster context, is a person's or group's capacity to anticipate, cope with, resist, and recover from the impact of a natural hazard" (Fothergill & Peek, 2004, p. 90). Cannon, as cited in Haque and Etkin (2007), "finds that the major determinants that make people vulnerable (i.e., social, economic and political factors), which determine the level of resilience of people's livelihoods and their ability to withstand and prepare for hazards are rarely tackled" (p. 276). Environmental disasters do not indiscriminately distribute risk and vulnerability or eliminate preexisting systems of racial and social stratification (Fothergill & Peek, 2004). In fact, these three environmental disasters reveal that people of color and poor people in the United States, U.S.-occupied territories, and around the world are disproportionately and differentially vulnerable because of preexisting systems of stratification. Political ecology research acknowledges the increased marginality and vulnerability of those who are socioeconomically disadvantaged and in particular how poverty, race, and gender intertwine and link with increased exposure to locally unwanted land uses (LULUs) which generally impose negative externalities to local neighborhoods and pose threats to their surrounding in the form of negative impacts as exemplified in the cases of Vieques, Katrina, and the Asian Tsunami (Bryant, 1998). "At issue here is how being poor or disadvantaged affects one's experiences in a disaster, from risk perception, to the post-disaster reconstruction of lives and communities" (Fothergill & Peek, 2004, p. 90). Thus, the environment is a politicized landscape reflecting "every day," "episodic," and "systemic" dimensions such as those highlighted in Table 1 (Bryant, 1998, p. 84).

Table 1: Dimensions of a Politicized Environment

Dimensions	Physical changes	Nature of human impact	Political response	Key concept
Everyday	Soil erosion, deforestation, salinization	Cumulative and typically highly unequal; the poor are the main losers	Livelihood protests/ resistance	Marginality
Episodic	Flooding, high winds/storms, drought	May have general impact but unequal exposure means that the poor are the main losers	"Disaster" relief	Vulnerability

Systemic	Nuclear fallout, pesticide concentration, biological species	Tends to have a general impact	Popular distrust of official "experts"	Risk

Central to this conceptualization of the environment as politicized is the recognition that the environment is constituted through struggles over material practices and over meaning that seek to equip all people and communities with the knowledge to mitigate, improve coping mechanisms, respond effectively, and recover with resiliency against environmental extremes. Yet as Vieques, Katrina, and the Asian Tsunami amply demonstrate, when faced with a social disaster, a disaster predicated upon and exacerbated by structural inequality and human decision making, who recovers is rooted in the choices that society makes and the prioritization of some lives over others. Thus, framing my argument through the lens of political ecology and critical race theory, I contend that social capital is a key factor in moving toward a culture of disaster prevention and risk reduction and that social capital can generate both the conditions necessary for mutual support and care and the mechanisms required for communities and groups to exert effective pressure to influence public policy. This article aims to demonstrate that disaster risk reduction begins at school, conclude with recommendations for policy and programming, and identify key elements needed to build social capital.

Portes (1998) maintains that "social capital stands for the ability of actors to secure benefits by virtue of membership in social networks or other structures" (p. 3). Though they should, all social networks or structures do not guarantee its participants access to social capital. Norris, Stevens, Pfefferbaum, Wyche, and Pfefferbaum (2008) emphasizes that genuine social capital derives from network structures and linkages, social support, sense of community, place attachment, citizen participation, and communication. Within this typology, social capital is translatable to the processes of resiliency and capacity building as well as participatory approaches that mobilize communities to prevent and mitigate disasters. In an analysis of evacuation decisions before Hurricanes Hugo and Andrew, it was found that residents with a stronger social support were twice as likely to evacuate as were residents with weaker social support (Norris et al.,

2008). ForPridmore, Thomas, Havemann, Sapag, and Wood (2007), communities with high levels of social capital are characterized by ethical homogeneity, strong leadership that is concerned with the welfare of the whole community, a high degree of trust, a sense of reciprocity, and mutual support leading to wide participation. Social capital, viewed this way, operates from the micro- (individual/family), meso- (neighborhood), city, and macro level (national/global). At the micro level, social capital operates through strengthening social support for individuals and vulnerable households; at the meso- and city level, social capital operates through capacity building that can lead neighborhood committees and community-based organizations (CBOs) participating with local government in proactively addressing local needs; and at the macro level, social capital operates through coalitions of CBOs, nongovernmental organizations (NGOs), and other civil society groups that mobilize community groups for a more just and sustainable world through change in policy and access to infrastructure and resources (Pridmore et al., 2007). These depictions are represented in Figure 1.

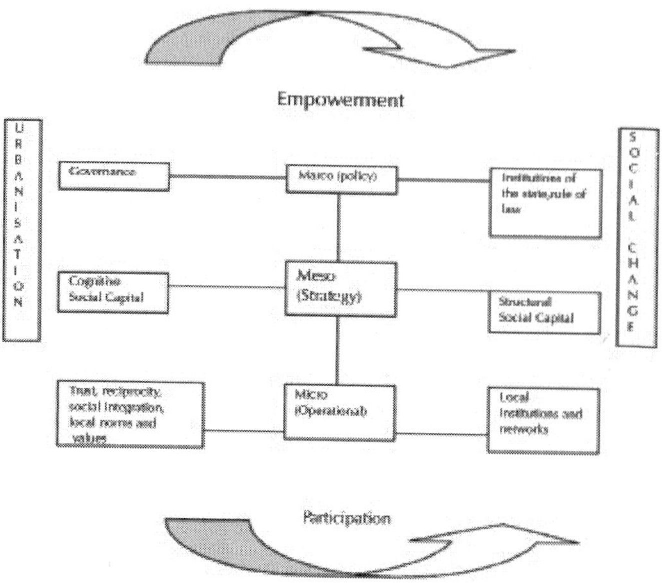

Figure 1: Types of social capital related to different levels and linked to participation and empowerment.

Evidence at the macro-, meso-, and micro level in considering the role of social capital operating in postdisaster response and recovery is revealed through survey, focus group, and interview data of Leong et al. (2007) that examined the lives of Southeast Asian Americans as compared with and contrasted with African Americans who were dislocated by the flooding in eastern New Orleans post Hurricane Katrina. Findings suggested that the Vietnamese American community benefitted from greater community capacity/social embeddedness, the advantages members received from their social ties (received and perceived) as well as the three key social-psychological dimensions of social capital (sense of community, place attachment, and citizen participation; Norris et al., 2008). In fact, filmmaker Spike Lee made a similar observation when commenting on Hurricane Katrina and the relief effort when he observed that during the rescue effort, it became evident that most of the people who sought refuge in the Louisiana Superdome or who perched on rooftops awaiting rescue were Black. Lee also proclaimed that what happened in New Orleans should be treated as a criminal act.

Conceptualized this way, Lee's assertion cannot simply be interpreted through the lens of American meritocracy which celebrates Asian Americans, like the Vietnamese American community in eastern New Orleans, as the "model minority" (Wu, 2002). It is more aptly construed through an analysis of the social capital resources available to the Vietnamese American community—linguistic, resistant, aspirational, familial, navigational, and societal (Leong et al., 2007) at the time Hurricane Katrina struck which contributed to a more vigorous disaster mitigation and preparedness response that was not available to the African American community. Such an analysis suggests to me that if anything is deserving of criminalization, it is the hidden curriculum that operated then and continues to operate at the macro, meso, and micro levels, which prevents disaster risk reduction from being taught in school to all populations and hierarchicalizes which populations are most vulnerable. According to Ivan Illich (1970), the hidden curriculum "schools" us and is at the root of extinguishing social imagination and assigning social roles void of social capital. Specifically, Illich states, the hidden curriculum of family life, draft, health care, so-called professionalism, or of the media play an important part in the institutional manipulation of man's world—vision, language, and demands . . . Each of us is personally responsible for his or her own deschooling, and only we have the power to do it. (p. 47)

Prevention begins with deschooling ourselves. Awareness is the first step toward action. Awareness can trigger interest, interest can lead to attention, and attention can prompt action. As schools are the best conduit for disseminating collective values, students and teachers can serve as vehicles for building a culture of prevention. Research (International Strategy for Disaster Reduction [ISDR], 2007) reveals diverse ways awareness has been raised among school students, teachers, and their communities in disaster-prone regions of the world such as Bangladesh, Indonesia, Sri Lanka, and Thailand. In these regions of the world, diverse initiatives include training of teachers, bringing disaster risk reduction into the classroom, organizing disaster curricula around disaster risk reduction knowledge, campaigning for disaster safety, and turning school students into catalysts and initiators.

For example, in 2005, in response to the December 2004 tsunami, ActionAid, in partnership with the Sustainable Development Resource Center (SDRC) funded the development of a kit for schoolchildren in the Bangla language that has been distributed with the aid of the Disaster Management Bureau (DMB) of the Bangladesh Government under the campaign "Disaster Risk Reduction Begins at School." As a part of the campaign, ActionAid launched a participatory and interactive approach to enable children to play a part in the community under the theme "Know Risk = No Risk." Since early 2006, the campaign has sought to build local capacity and awareness through (a) development and enhancement of learning tools, (b) training of facilitators and teachers, (c) participatory extracurricular activities, (d) testing children's cognitive response, (e) drawing lessons for scaling up, and (f) school and community workshops. Most importantly, since the introduction of this kit in 2006, by ActionAid partners in 26 primary and secondary schools, youth and teachers have been engaged as disaster risk reduction practitioners by being trained to (a) understand and identify risks and communicate these to others, and explore ways to raise community awareness; (b) engage in participatory hazard risk mapping of their surroundings, community, village, and so on; and (c) engage in family and school risk reduction planning and management (International Strategy for Disaster Reduction, 2007). As another illustration, the December 2004 tsunami in Aceh Province was a catalyst that caused the Indonesian government to take note that children and women are most vulnerable to disasters. In light of this reality and given that a large segment of Indonesian society lacks understanding of hazards and disasters, the Indonesian government decided to equip primary students with some knowledge of natural hazards and risk reduction measures. The project, titled "Disaster Awareness in Primary Schools" (DAPS), was launched in high-risk primary schools in Indonesia in 2005 and is ongoing. The DAPS project first aimed to secure recognition and understanding of natural hazards, including the potential impacts these can have on personal environments and social development. The project also aimed to develop disaster prevention and mitigation knowledge and skills among families, schools, and communities, and knowledge of disaster response measures for extreme situations. The project approach is highlighted in Figure 2.

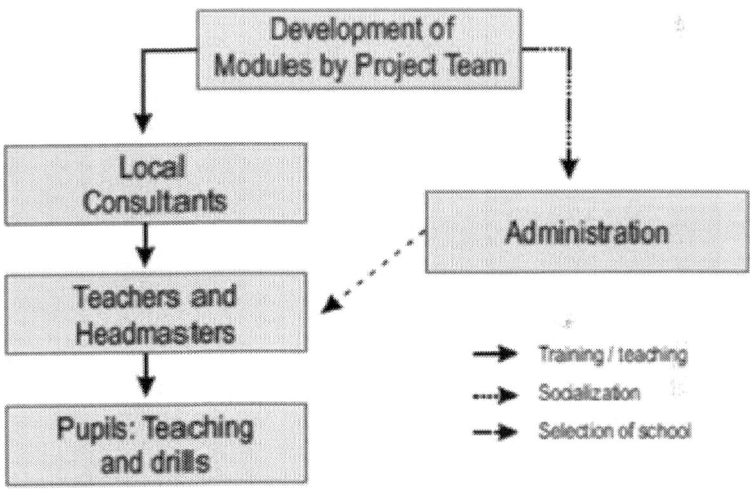

Figure 2: Disaster Awareness in Primary Schools approach.

The ISDR and United Nations has determined DAPS to be a model project for teaching disaster risk reduction in schools because it takes teachers' knowledge into account in the following ways: (a) It enforces clarity, precision, and perseverance by committing the project team to the rules "Simplify but do not falsify!" and "Don't hope it works, make sure it works!" (b) Information had to be noted down in great detail; (c) a learning by doing approach was applied in engaging participants in practical activities like first-aid courses; and (d) traditional approaches in writing training modules were abandoned in favor of narrative form. As a result, both students and teachers widened their knowledge base and a pupil-centered methodology was adopted (ISDR, 2007). Because evidence indicates that from an early age, young people are capable of insightful and constructive analysis of their experiences of learning in school (Delpit, 1988; Greene, 1988; Illich, 1970; Pedder & McIntyre, 2006; SooHoo, 1993), these campaigns and projects reveal the potential in consulting students as purveyors of knowledge in disaster prevention and reduction.

This evidence bears fruit in the example of Sri Lanka that, like Bangladesh and Indonesia, learned from the December 2004 tsunami and utilized it as a turning point in the history of disaster mitigation to prompt relevant stakeholders to collaborate toward a long-term, comprehensive, and holistic disaster risk management framework with a mitigation-based approach. Sri Lanka's Disaster Management Act No. 13 of 2005 was developed, providing a solid legislative and institutional arrangement for disaster risk management, which led to the establishment of both the National Council for Disaster Management—under the President of the Republic—and the Disaster Management Center in May 2005, as well as the creation of the Ministry of Disaster Management in November 2005. Recognizing the importance of education in this area, the Government of Sri Lanka has sought to integrate disaster risk management into the national education system. In the spirit of this mandate, teachers have been trained in India on this topic and have helped develop a teacher training curriculum in Sri Lanka which has helped to boost "disaster safety education" in the country (ISDR, 2007, pp. 33-36). In Thailand, disaster risk reduction also became a focal point for child-focused NGOs and aid agencies after the December 2004 Asian Tsunami. As children are known to be the most vulnerable to disasters, Save the Children, a child-focused NGO, initiated in 2006 a school project called "Child-Led Disaster Risk Reduction (CLDRR) in Thailand Project." The rationale behind the project which is still in place is that children and youth can play an active role in community affairs that are relative to them (ISDR, 2007).

Such shifts and implementations in the American disaster prevention culture post Hurricane Katrina at the macro (national/global), meso (structural), and city level, like that which has occurred in Bangladesh, Indonesia, Sri Lanka, and Thailand post-Asian Tsunami, would equip schools with the curricular tools to train students as disaster risk reduction practitioners irrespective of the linguistic, resistant, aspirational, familial, navigational, and societal capital of their respective communities. In the Vietnamese community in New Orleans post Hurricane Katrina, their minimal reliance on institutional care along with perceptions of themselves as socially powerful compelled the community to use their social capital to maneuver. Subsequently, because of the network structures and linkages, social support, and social capital possessed by the Vietnamese American community, they evaded the ethos of the hidden curriculum operating in the aftermath of Hurricane Katrina at the macro-, meso-, and city level. Such tools enabled the community to tap into immigrant and pan-ethnic Asian American networks both locally and nationally. Familial capital was evident in the preexisting bonds shared among residents within the Versailles and Village d'l Est Vietnamese American community who immigrated to eastern New Orleans from the same village in Vietnam. Accordingly, the Vietnamese community was able to mobilize this capital in postdisaster evacuation, recovery, and resilience. Resistant capital emerged from the historical specificity of the Vietnamese immigrants in New Orleans as political refugees who were thrust into rebuilding their lives after immigrating to the United States. Navigational capital was exhibited by second-generation Vietnamese American children who, because of their bilingual ability, were savvy enough to navigate bureaucracy related to education, housing, assistance, and medical care in the United State for themselves and their parents which proved indispensable in navigating government assistance, moving into and out of shelters, and locating housing post Katrina (Leong et al., 2007).

It should also be remembered that of the 55% of the prestorm population of New Orleans that were renters, a significant number were African American (U.S. Census Bureau, 2007). While this community possesses a semblance of social, familial, and aspirational capital in their desire to return home, the lack of navigational capital and the extortionate rental market remain key impediments. Because of this, many African American survivors of Hurricane Katrina, who remained or returned, were relegated to residing in formaldehyde-laden Federal Emergency Management Agency (FEMA) trailers that have since been determined by the Centers for Diseases Control to predispose them to a variety of respiratory ailments, some of which have resulted in unnecessary deaths (Centers for Disease Control and Prevention, 2008). Although residents may have received rental vouchers for temporary housing, housing from rent was in very short supply. In the Upper Ninth Ward, 87% of rental housing was flooded and in the Lower Ninth Ward, 75% of rental housing was flooded. Rent levels rose precipitously with such short supply of rental units. One indicator of rent level is the HUD Fair Market Rent (FMR) standard, which is set at the 40th percentile of rents for the metropolitan area (FMR also dictates the upper bounds of acceptable voucher rents). FMRs have risen by approximately 45 percent since the autumn of 2005, with the FMR for a one bedroom unit rising USD 578 in 2005 to USD 836 in 2007. The lack of housing available for rent, especially at affordable levels, has become an impediment to reconstruction in the low to moderate income neighborhoods of the Ninth Ward. (Green, Bates, & Smyth, 2007, p. 326)

"The federal 'Road Home' program, which is intended to provide cash assistance to those whose homes were damaged by the storm, serves only homeowners and not renters" (Katz, 2008, p. 22). What is more, in exploring the postdisaster experience of Latinos in New Orleans, Trujillo-Pagán (2007) explains that their experience reflected their social and political vulnerability as undocumented immigrants and found that they experienced Hurricane Katrina as a racial/ethnic minority as they faced a variety of structural obstacles in accessing relief services, including ineligibility and discrimination. Yet social and navigational capital alone is not sufficient enough an explanation.

Based on the analysis of Harris-Lacewell (2007), which utilized a simple regression model to juxtapose the September 11 tragedy with that of Hurricane Katrina to examine the racial disparity in emotional response to Katrina between Black and White Americans, the fact that Black Americans viewed Katrina as a racialized disaster, while White Americans did not, was not based on psychology but rather on the historical and contemporary racial beliefs that shape the political and emotional lives of African Americans. Racial inequality has persisted into the 21st century. African Americans continue to earn a fraction of their white counterparts, suffer from worse physical health, enjoy fewer educational opportunities, and are less well represented in politics [the 2008 election of President Barack Obama being the exception] and popular culture, and labor under persistent racial stereotypes. (Harris-Lacewell, 2007, p. 34)

Thus, the symbolism of the broken levees and the destruction of New Orleans as a city are laden with disillusionment about the contemporary state of African American citizenship. This feeling of second-class citizenship is rooted in a history of colonialism. Colonialism is the act of a mother-nation assuming political and/or economic control over another, whereby the mother-nation defines the colony's laws and general administration in addition to acquiring a significant portion of the colony's generated revenues. The Louisiana area was colonized by the French, the Spanish, and finally the United States as a result of the Louisiana Purchase. Each of these colonizing forces brought with them its own definitions and understandings of race and ethnicity. Thus, the status of people of color in this region varied depending upon the colonizing government. (Miller & Rivera, 2007, p. 24)

In the south, colonialism, as a system of racial and social stratification, evolved during the antebellum era, Reconstruction, the Civil War era, and embedded itself in Jim Crow segregation. Therefore, the preparations for Hurricane Katrina, and the response afterward, followed the historic pattern of class and racial divides and differential responses; it showed that the intersection of race and class continues to influence the vulnerability of disadvantaged groups to future disasters and clearly influences their capacity to recover. (Cutter & Smith, 2009, p. 31)

The colonial past that unambiguously and systematically impacts contemporary life in the Gulf Coast states, like Louisiana, and undoubtedly influences recovery and reconstruction in New Orleans also shapes the island of Vieques efforts to recover and reconstruct its land after the departure of the U.S. Navy Atlantic Fleet May 1, 2003. Although Vieques is a political municipality of Puerto Rico with its own mayor, between 1941 and 2003, the U.S. Navy Atlantic Fleet owned and managed more land than the municipality itself. Since May 1, 2003, the U.S. Department of the Interior has inherited the position of primary landholder of Vieques and subsequently designated it as the Vieques National Wildlife Refuge. The Vieques National Wildlife Refuge now constitutes the largest wildlife refuge in the Caribbean. The navy was so fond of Vieques because it afforded them the opportunity to practice beach landings, special operations parachute drops, and small-arms target practice in the maneuver area. By their accounts, they could create realistic and integrated scenarios that simulated true combat coordination among different units and countries. Moreover, Vieques occupied a strategic location, with weapons storage space close to Cuba and the Panama Canal (Duchesne Winter, 2007). Although the island was managed by the navy, it was accessible to other military branches that required a practice range as they prepped for combat in the Middle East and the Balkans. Therefore, today Vieques, a poor community of color, stands, like the levees of New Orleans, as emblematic of historical malfeasance and exploitation orchestrated by the U.S. Navy for purposes of securing a landscape for waste disposal and/or experimentation with risky technologies (toxic colonialism). The military created several dump sites (LULUs) characterized by inadequate initial site preparation causing pollutants to readily leach into soil and water. Solvents, diesel, fuel, solid waste, and even weapons themselves were thrown into these dumps. There are two known submarine disposal sites in the waters of the Southeast where the contents of photographed barrels remain. Coral reef destruction has also been documented. So synthetic chemicals continue to precipitate on western side of the island and heavy metals are predominant on the eastern side of the island. Presumably on the western side of the island, the military used industrial solvents to clean vehicles and equipment and used lead-based paints (Arbona, 2004). Evidence of environmental racism, the inequitable distribution of environmental hazards based on race, can be seen in lead poisoning in children. In fact, the Federal Agency for Toxic Substances and Disease Registry considers lead poisoning to be the number one environmental health problem for children in the United States (Centers for Disease Control and Prevention, 2005). Furthermore, during the 62 years of military activities, airborne pollutants and pulverized rock particles reached the civilian population through the strong East-to-West Caribbean trade winds, while people might have ingested other chemicals through pollution from aquifers and drinking wells as well as ingestion of plants and animals with high levels of heavy metals. Besides, Dr. Arturo Massol and Elba Diaz have documented high levels of copper, zinc, nickel, cobalt, and cadmium in crustaceans by demonstrating that crustaceans on the bombing range had levels of these metals up to 80 times higher than typical market samples. In the civilian areas, they also found that the crustaceans had levels between 10 and 20 times higher than typical market samples. Experimentation on plants and vegetables revealed high levels of lead, cobalt, and manganese in plants and also found lead and cadmium in pumpkins, peas, pineapples, and yucca (indigenous plants). Massol and Diaz concluded that these heavy metals, originating at the bombing range, continue to travel through the food chain and impact the local diet and health of the Viequenses (Arbona, 2004). So, a central issue that continues to confront Vieques in 2009 is the slow pace of federal property transfer and the commitment of funds for cleanup because Puerto Rico lacks representation in the federal government (McCaffrey, 2002). Colonialism, racism, and socioeconomic marginalization in the Vieques crisis, as in the Katrina crisis, continue to compound difficulties in ensuring that the federal government meets its obligations.

Still, 3 years after Hurricane Katrina, the response to Hurricane Ike of 2008, which struck Galveston, reflect disparities in the ways in which race and class are treated along America's hurricane coasts. As a result, most hurricane evacuations have reflected the suburban experience: majority white residents with the resources and private transportation to do so temporarily evacuate the coast, and after staying with family or friends in motels, quickly return to begin the cleanup. (Cutter & Smith, 2009, p. 33)

Although there were several fatalities, mostly White middle-income residents, in Galveston resulting from Hurricane Ike, many residents on the Texas coast determined that the risks of a Category 2 storm did not warrant evacuation and misunderstood that Ike's storm surge as synonymous with a Category 4 storm such that social capital disparity is reflected in the evacuation experiences during the 2008 hurricane season (Cutter & Smith, 2009). The residents of Galveston were given a choice to evacuate or stay and the residents of New Orleans were not. Research following Hurricane Andrew is also instructive in that it established some relationships between income and preparedness revealing that poorer people could not afford flood insurance even thought they were aware of its availability and benefits. Furthermore, it revealed that socioeconomic status impacts who receives formal warning signals (emergency broadcasts and flood sirens) of an impending danger and heeds recommendations to evacuate (Fothergill & Peek, 2004). Hurricane Katrina demonstrated that although many heard the storm warnings, few were able to take action because they lacked money for transportation. It is known that many New Orleanians lacked cars. A 2000 census revealed that 27% of New Orleans households, about 120,000 people, were without transportation (Miller & Rivera,

2007). Clearly, socioeconomic status affected both the location of the community and access to necessary resources during this emergency. Poverty can explain the lack of access to fuel, water, and supplies, as well as the inability to evacuate due to a lack of private transportation. "This constellation of factors prevented many people from being able to evacuate on their own. Consequently, most of those stranded in the city were the poor, elderly, sick, and those caring for the sick" (Fox & Gibbons, 2005, p. 518). This is an issue that needs much attention, policy development, and stakeholder engagement.

Moreover, Fothergill and Peek (2004) maintain that not only does stakeholder engagement increase steadily with income level, but education also combats fatalism and thereby fosters more stakeholder engagement (citizen participation). It is through education that community members gain valuable knowledge and information which translates to power, power that restructures curricula that fosters the acquisition of skills that match immediate needs of rebuilding postecological disaster such as carpentry, masonry, and plumbing, but that can also be used after rebuilding is completed to expand livelihood options and thus enhance navigational and social capital. For example, in New Orleans, these restructured curricula might be taken up by charter schools, religious schools, and newly restored and reorganized public schools. Specifically, "citizen participation is the engagement of community members in formal organizations, including religious congregations, school and resident associations, neighborhood watches, and self-help groups" (Norris et al., 2008, p. 139). Resilience is "the capacity of a system, community or society potentially exposed to hazards to adapt, by resisting or changing in order to reach and maintain an acceptable level of functioning and structure" (O'Brien, 2006, p. 74). Citizen participation is strengthened when coupled with resilience as a process. Citizen participation was evident in the post-Katrina recovery and reconstruction of the Vietnamese American community in New Orleans. Thus, enhancing resilience increases stakeholder engagement at all levels. As cited by O'Brien (2006), David McEntire states, recognizing the human role in disasters, taking responsibility for action, having a disaster plan, building capabilities to implement the plan . . . and sharing information about recovery priorities are processes that can enhance resilience for an individual, group, community or nation to deal with unique destabilizing events. (p. 74)

Other representative definitions of resilience can be found in Table 2 (Norris et al., 2008, p. 129).

Table 2: Representative Definitions of Resilience

Citation first author, year	Level of analysis	Definition
Gordon, 1978	Physical	The ability to store strain energy and deflect elasticity under a load without breaking or being deformed
Bodin, 2004	Physical	The speed with which a system returns to equilibrium after displacement, irrespective of how many oscillations are required
Holling, 1973	Ecological system	The persistence of relationships within a system; a measure of the ability of systems to absorb changes of state variables, driving variables, and parameters, and still persist
Waller, 2001	Ecological system	Positive adaptation in response to adversity; it is not the absence of vulnerability, not an inherent characteristic, and not static
Klein, 2003	Ecological system	The ability of a system that has undergone stress to recover and return to its original state; more precisely (a) the amount of disturbance a system can absorb and still remain within the same state or domain of attraction and (b) the degree to which the system is capable of self-organization
Longstaff, 2005	Ecological system	The ability of an individual, group, or organization to continue its existence (or remain more or less stable) in the face of some sort of surprise . . . Resilience is found in systems that are highly adaptable (not locked into specific strategies) and have diverse resources
Resilience Alliance, 2006	Ecological system	The capacity of a system to absorb disturbance and reorganize while undergoing change so as to still retain essentially the same function, structure, and feedback—and therefore the same identity
Adger, 2000	Social	The ability of communities to withstand external shocks to their social infrastructure
Bruneau, 2003	Social	The ability of social units to mitigate hazards, contain the effects of disasters when they occur, and carry out recovery activities in ways that minimize social disruption and mitigate the effects of future earthquakes
Godschalk, 2003	City	A sustainable network of physical systems and human communities, capable of managing extreme events; during disaster, both must be able to survive and function under extreme stress
Brown, 1996	Community	The ability to recover from or adjust easily to misfortune or sustained life stress
Sonn, 1998	Community	The process through which mediating structures (schools, peer groups, family) and activity settings moderate the impact of oppressive systems
Paton, 2000	Community	The capability to bounce back and to use physical and economic resources effectively to aid recovery following exposure to hazards
Ganor, 2003	Community	The ability of individuals and communities to deal with a state of continuous, long-term stress; the ability to find unknown inner strengths and resources to cope effectively; the measure of adaptation and flexibility
Ahmed, 2004	Community	The development of material, physical, sociopolitical, sociocultural, and psychological resources that promote safety of residents and buffer adversity
Kimhi, 2004	Community	Individuals' sense of the ability of their own community to deal successfully with the ongoing political violence
Coles, 2004	Community	A community's capacities, skills, and knowledge that allow it to participate fully in recovery from disasters
Pfefferbaum, 2005	Community	The ability of community members to take meaningful, deliberate, collective action to remedy the impact of a problem, including the ability to interpret the environment, intervene, and move on
Masten, 1990	Individual	The process off, capacity for, or outcome of successful adaptation despite challenging or threatening circumstances
Egeland, 1993	Individual	The capacity for successful adaptation, positive functioning, or competence . . . despite high-risk status, chronic stress, or following prolonged or severe trauma
Butler, 2007	Individual	Good adaptation under extenuating circumstances; a recovery trajectory that returns to baseline functioning following a challenge

Arguably, if resilience is to be obtained and play a pivotal role in reducing disaster risk, then the dominant paradigm of disaster mitigation and planning must shift away from an outcome-focused paradigm to one that is process oriented. O'Brien (2006, p. 75) offers a framework for this new paradigm in Table 3.

Table 3: Resilience and Disaster Management

Dominant paradigm	**New paradigm**
Isolated event	Part of development
Risk not normal	Risk of disaster/conflict
Techno-lega	Social capacity
Centralized	Participatory
Low accountability	Transparent
Postevent planning	Predisaster plans
Status quo restored	Transformation
Cross-Cutting Theme: Public Education	

Although the terminology on the left-hand column of Table 3 currently characterizes the U.S. approach, in the case of New Orleans in the aftermath of Hurricane Katrina, it is irrefutable that the future of New Orleans is predicated upon embracing a new paradigm of transformation. It is clear that public education must be the vehicle to foster this paradigm change because education is at the heart of social reproduction and schools anchor communities. "Media also can be engaged to publicize available services and educate the public about typical reactions to disaster" (Norris et al., 2008, p. 140). In addition, in events such as hurricanes and other ecological disasters, an error proof risk communication and crisis communication plan needs to be in place. It is clear that such a communication plan was lacking in Hurricane Katrina and the Asian Tsunami. "While crisis communication aims to reduce negative outcomes associated with a crisis, risk communication messages are intended to encourage the receiver to engage in proactive behaviors ahead of time to avoid threats and negative consequences" (Lachlan, Spence, & Eith, 2007, p. 204). Communication, crisis, and risk linked with social capital lead to human agency (community competence) of which collective action and decision making are components. Articulated this way, multiple pathways exist that lead communities toward resilience in the face of incomprehensible disasters; they are summarized in a framework in Figure 3 (Norris et al., 2008, p. 136)

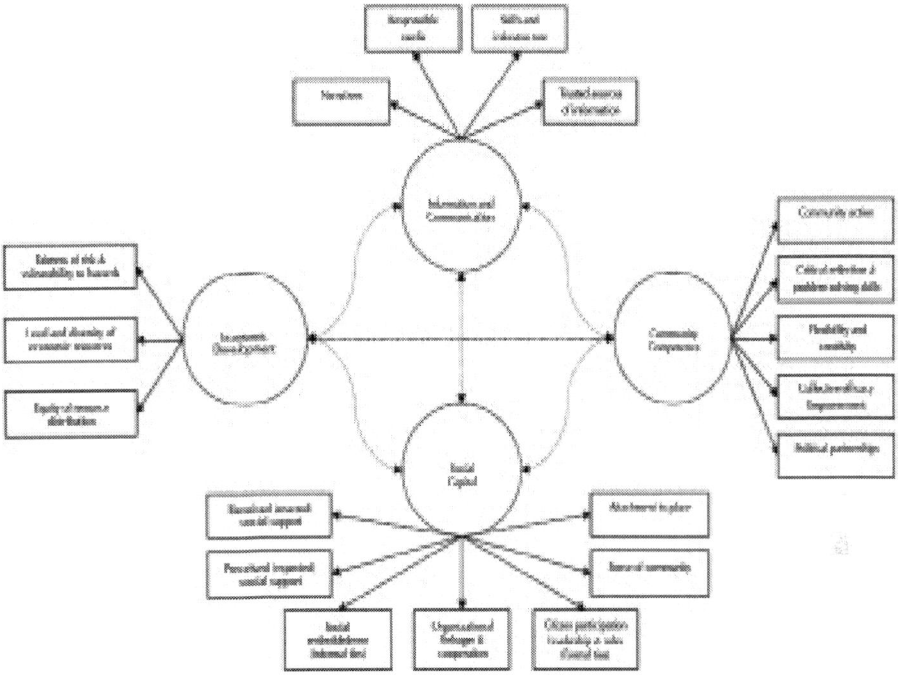

Figure 3: Community resilience as a set of networked adaptive capacities.

As feminist historian Gerda Lerner (1986) observes, if we believe patriarchy is the natural state of affairs, we then have no hope of creating a new paradigm or an alternate system. "The disaster of Hurricane Katrina offers a basic tenet for future emergency preparedness planning: when technology and political processes break down, it comes down to human relationships" (Leong et al., 2007, p. 182). Based on the discussion in this article, a checklist of 10 key elements emerges that are prerequisites for building social capital as part of a social justice process. These elements are listed in Table 4 (Pridmore et al., 2007, p. 139).

Table 4: Ten Actions Needed to Build Social Capital as Part of a Social Justice Process

Actions
1. Assessing the context and asking the right questions: The choice of questions is influenced by the expected size and direction of disaster reduction and prevention impacts, the prominence of the issue in the government's policy agenda, and the timing and urgency of the underlying disaster mitigation and preparedness policy or strategy.
2. Identifying stakeholders: Stakeholder analysis identifies the people, groups, and organizations that are important to consider when looking at disaster reduction and prevention impacts.
3. Developing the capacity of stakeholders to take action and build social capital and cohesion: The expected policy change can only take place if sufficient knowledge, skills, and resources are in place.
4. Assessing institutions and creating opportunity to ensure intersectoral collaboration: Institutions determine the framework in which policy reforms may affect stakeholders in government, private sector, and civil society, and are the main arenas in which stakeholders interact with one another.
5. Strengthening the demand side of governance: Assessing and ensuring people's participation from the organizational and legal aspect, including a concern with ensuring access to data.
6. Strengthening institutions' role, function, and structure: Involves organizing and creating critical links between the policy objectives, policy actions, and their impacts on key stakeholder groups within the health and other sectors at various levels.
7. Mobilizing resources: To the extent that they are necessary for social change.
8. Advocate for upscaling and change: Policy and advocacy to relevant stakeholders at different levels.
9. Monitoring and evaluating impacts: Provides an opportunity to set up early stage systems for monitoring.
10. Identifying the appropriate level and type of intervention: Individual, neighborhood, city, etc.

These 10 key elements are important in building social capital and provide a useful framework, along with adopting new/alternative paradigms that are not complicitous with current economic, education, and political structures, for policy development and practice. Furthermore, they help to realize the contributions that social capital can make toward disaster mitigation and preparedness in a globalized world. Through the lens of social capital, disaster mitigation and preparedness might best be taken up through disaster education integrated into K-12 curricula. Though disaster education has not been fully implemented in K-12 curricula, the theme of "Disaster Reduction, Education, and Youth" is not new having been introduced during the UN World Disaster Reduction Campaign in 2005 as a priority integral to the 2005-2015 Hyogo Framework for Action as part of Priority 3, focusing on the "use [of] knowledge, innovation, and education to build a culture of safety and resilience at all levels" ((UNISDR, 2005, p. 53). Accordingly, social actors (activists connected with NGOs who have worked with dedication, empathy, and coalitional consciousness to bring this priority to life at both grassroots and policy levels) are essential to framing this informal and formal education framework, via environmental education, capable of taking on the challenge of disaster risk reduction education.

REFERENCES

1. Arbona, J. (2004). *Vieques, Puerto Rico: From devastation to conservation, and back again* (Master's thesis). Retrieved from Massachusetts Institute of Technology.
2. Bankoff, G. (2001). Rendering the world unsafe: "Vulnerability" as Western discourse. *Disasters, 25*, 19-35.
3. Bryant, R. L. (1998). Power, knowledge and political ecology in the third world: A review. *Progress in Physical Geography, 22*, 79-94.
4. Centers for Disease Control and Prevention. (2005). *CDC's third national report on human exposure to environmental chemicals: Spotlight on lead*. Retrieved from http://www.cdc.gov/exposurereport/pdf/factsheet_lead.pdf
5. Centers for Disease Control and Prevention. (2008). *CDC releases results of formaldehyde level tests: FEMA to expedite relocation of residents from temporary housing units*. Retrieved from http://www.cdc.gov/media/pressrel/2008/r080214b.htm
6. Cutter, S., & Smith, M. (2009). Fleeing from the hurricane's wrath: Evacuation and the two Americas. *Environment, 51*(21), 26-39.
7. Delpit, L. (1988). The silenced dialogue: Power and pedagogy in educating other people's children. *Harvard Educational Review, 58*, 280-298.
8. Duchesne winter, J. (2007). Vieques: Protest as a consensual spectacle. In F. Negrón-Muntaner (Ed.), *None of the above: Puerto Ricans in the global era* (pp. 87-97). New York, NY: Macmillan.
9. Fothergill, A., & Peek, L. A. (2004). Poverty and disasters in the United States: A review of recent sociological findings. *Natural Hazards, 32*, 89-110.
10. Fox, C. E., & Gibbons, M. C. (2005). Disaster response: One size does not fill all. *Journal of Urban Health: Bulletin of New York Academy of Medicine, 82*, 517-519.
11. Green, R., Bates, L. K., & Smyth, A. (2007). Impediments to recovery in New Orleans' upper and lower ninth ward: One year after hurricane Katrina. *Disasters, 31*, 311-335.
12. Greene, M. (1988). *The dialectic of freedom*. New York, NY: Teachers College Press.

13. Haque, C. E., & Etkin, D. (2007). People and community of constituent parts of hazards: The significance of societal dimensions in hazards analysis. *Natural Hazards, 41*, 271-282.

14. Harris-Lacewell, M. (2007). Do you know what it means: Mapping emotion in the aftermath of Katrina? *Souls: A Critical Journal of Black Politics, Culture, and Society, 9*, 28-44.

15. Illich, I. (1970). *Deschooling society*. London, England: Marion Boyars.

16. International Strategy for Disaster Reduction. (2007, November). *Towards a culture of prevention: Disaster risk reduction begins at school: Good practices and lessons learned*. Presented at United Nations Proceedings, Geneva, Switzerland.

17. Katz, C. (2008). Bad elements: Katrina and the scoured landscape of social reproduction. *Gender, Place & Culture, 15*, 15-29.

18. Lachlan, K., Spence, P. R., & Eith, C. (2007). Access to mediated emergency messages: Differences in crisis knowledge across age, race, and socioeconomic status. In K. A. Bates & R. S. Swan (Eds.), *Through the eye of Katrina: Social justice in the United States* (pp. 203-219). Durham, NC: Carolina Academic Press.

19. Leong, K. J., Airriess, C., Chia-Chen Chen, A., Keith, V., Li, W., Wang, Y., & Adams, K. (2007). From invisibility to hypervisibility: The complexity of race, survival, and resiliency for the Vietnamese-American community in Eastern New Orleans. In K. A. Bates & R. S. Swan (Eds.), *Through the eye of Katrina: Social justice in the United States* (pp. 169-185). Durham, NC: Carolina Academic Press.

20. Lerner, G. (1986). *The creation of patriarchy*. New York, NY: Oxford University Press.

21. McCaffrey, K. T. (2002). *Military power and popular protest: The U.S. navy in Vieques, Puerto Rico*. New Brunswick, NJ: Rutgers University Press.

22. Miller, D. S., & Rivera, J. D. (2007). Setting the stage: Roots of social inequity and the human tragedy of hurricane Katrina. In K. A. Bates & R. S. Swan (Eds.), *Through the eye of Katrina: Social justice in the United States* (pp. 15-35). Durham, NC: Carolina Academic Press.

23. Norris, F. H., Stevens, S. P., Pfefferbaum, B., Wyche, K. F., & Pfefferbaum, R. L. (2008). Community resilience as a metaphor, theory, set of capacities, and strategy of disaster readiness. *American Journal of Community Psychology, 41*, 127-150.

24. O'Brien, G. (2006). UK emergency preparedness: A step in the right direction. *Journal of International Affairs, 59*(2), 63-87.

25. Pedder, D., & McIntyre, D. (2006). Pupil consultation: The importance of social capital. *Educational Review, 58*, 145-157.

26. Portes, A. (1998). Social capital: Its origins and applications in modern sociology. *Annual Review of Sociology, 24*, 1-24.

27. Pridmore, P., Thomas, L., Havemann, K., Sapag, J., & Wood, L. (2007). Social capital and healthy urbanization in a globalized world. *Journal of Urban Health: Bulletin of the New York Academy of Medicine, 84*, 130-143.

28. SooHoo, S. (1993). Students as partners in research and restructuring schools. *Educational Forum, 57*, 386-393.

29. Trujillo-Pagán, N. (2007). Katrina's Latinos: Vulnerability and disasters in relief and recovery. In K. A. Bates & R. S. Swan (Eds.), *Through the eye of Katrina: Social justice in the United States* (pp. 147-167). Durham, NC: Carolina Academic Press.

30. United Nations International Strategy for Disaster Reduction. (2005). *Disaster through a different lens: Behind every effect, there is a cause*. Geneva, Switzerland: Author.

31. U.S. Census Bureau. (2007). *New Orleans (city) quick facts from the US Census Bureau*. Retrieved from http://quickfacts.census.gov/qfd/states/22/2255000.html

32. Wu, F. H. (2002). The model minority: Asian American "success" as a race relations failure. In F. H. Wu (Ed.), *Yellow: Race in America beyond black and white* (pp. 39-77). New York, NY: Basic Books.

Chapter 2

Factors Influencing Readiness to Deploy in Disaster Response: Findings from a Cross-sectional Survey of the Department of Veterans Affairs Disaster Emergency Medical Personnel System

Nicole K Zagelbaum[1], Kevin C Heslin[1], Judith A Stein[1], Josef Ruzek[2], Robert E Smith[3], Tam Nyugen[2], and Aram Dobalian[1, 4]

[1]Veterans Emergency Management Evaluation Center, 16111 Plummer Street MS-152, North Hills, CA 91343, USA

[2]National Center for PTSD, 795 Willow Rd, Menlo Park, CA 94025, USA

[3]Veterans Health Administration Office of Emergency Management, 510 Butler Ave Bldg. 203B, Martinsburg, VA 25405, USA

[4]Department of Health Policy and Management, University of California, Los Angeles Fielding School of Public Health, 650 Charles Young Dr. S., Los Angeles, CA 90095, USA

ABSTRACT

Background

The Disaster Emergency Medical Personnel System (DEMPS) program provides a system of volunteers whereby active or retired Department of Veterans Affairs (VA) personnel can register to be deployed to support other VA facilities or the nation during national emergencies or disasters. Both early and ongoing volunteer training is required to participate.

Methods

This study aims to identify factors that impact willingness to deploy in the event of an emergency. This analysis was based on responses from 2,385 survey respondents (response rate, 29%). Latent variable path models were developed and tested using the EQS structural equations modeling program. Background demographic variables of education, age, minority ethnicity, and female gender were used as predictors of intervening latent variables of *DEMPS Volunteer Experience*, *Positive Attitude about Training*, and *Stress*. The model had acceptable fit statistics, and all three intermediate latent variables significantly predicted the outcome latent variable *Readiness to Deploy*.

Results

DEMPS Volunteer Experience and a *Positive Attitude about Training* were associated with *Readiness to Deploy*. *Stress* was associated with decreased *Readiness to Deploy*. Female gender was negatively correlated with *Readiness to Deploy*; however, there was an indirect relationship between female gender and *Readiness to Deploy* through *Positive Attitude about Training*.

Conclusions

These findings suggest that volunteer emergency management response programs such as DEMPS should consider how best to address the factors that may make women less ready to deploy than men in order to ensure adequate gender representation among emergency responders. The findings underscore the importance of training opportunities to ensure that gender-sensitive support is a strong component of emergency response, and may apply to other emergency response programs such as the Medical Reserve Corps and the American Red Cross.

BACKGROUND

Organizational training programs for emergency responders are critical to quickly identify and deploy appropriate resources in the overall response effort [1]. Results from several studies suggest that training is linked with greater effectiveness of emergency response efforts and that training of health personnel is positively associated with willingness to volunteer and deploy in the event of a disaster [2-4]. However, there exist several significant gaps in training for emergency providers. For example, an Institute of Medicine report found that only 49% of hospitals trained their residents and interns in disaster preparedness, which represents a national point of concern in the emergency management community [5]. A number of studies suggest that willingness to report to work could be a substantial problem within the disaster response workforce, depending on the type of event. Several large surveys have found that the percent of healthcare workers willing to report to work during disasters and emergencies can be 50% or less for severe snow storms [6] and influenza pandemics [7,8], and less than 60% for chemical or radiation events [9], leading to potentially crippling personnel shortage in times of great need. It is important to recognize the factors that impact an appropriate medical response to a disaster in a global context. A survey of over 900 nurses in China found that as few as one-third were willing to report to work during an infectious disease outbreak and also found that the most significant factor affecting to willingness to work was clinical experience [10]. Another survey of primary care health staff in Guangdong Province, China found that only one-fourth of respondents had participated in emergency response in the past, reflecting both poor response capacity and lack of experience [11]. In a survey of Israeli nurses, less than half had reported to work when asked in the past year following a response [12]. Another Israeli survey indicated that only 51% of hospital workers said they would report following a missile attack [13]. Finally, in a survey of hospital workers in the United Kingdom, the average likelihood of reporting to work in the event of an influenza pandemic was 59.3%; rates varied by factors such as familial support and previous disaster training [14].

Underscoring the critical role of knowledge and skills in competent disaster response work, the World Association for Disaster and Emergency Medicine [15] and other organizations have recommended guidelines for education and training programs for disaster medicine and humanitarian aid. The need for these recommendations is supported by work from around the globe. For example, among Australian emergency prehospital health care providers, less than 5% of employees felt they had been adequately trained concerning avian influenza [16]. Another article from Australia reported that willingness to work during a pandemic was associated with increased knowledge, education, and training about infectious agents among healthcare workers [17].

Limited work exists on the role of gender in disaster response readiness – i.e. the ability and willingness of responders of both genders to deploy and perform assigned duties [18]. One survey of healthcare workers found that women were less willing to report to duty in the event of a disaster [6]. An evaluation of Veterans found that women report stress to a greater extent than do men following similar levels of exposure to deployment stressors [19]. Gender differences in psychological stress after disasters are also important to note among emergency responders, as this could impact the work performance of deployed personnel [20].

Disasters and public health emergencies increase the general demand not only for emergency services and basic health care, but also for a wide range of gender-specific health and social services [21]. For example, in the weeks after an 8.0-magnitude earthquake in China, women reported markedly increased symptoms of lower genital tract infection, pelvic inflammatory disease, and menstrual disorders [22]. After the 2001 attacks on the World Trade Center, women were more likely to report using psychiatric medications than were men [23]. Women responders may be able to communicate and interact with women with disaster-related distress or gender-specific health and social needs more effectively than male responders. For example, after responses by the police to domestic violence calls, female officers receive more positive evaluation than do male officers from battered women by providing increased empathy, referral services, and legal information [24,25]. In a survey of over 800 patients visiting the emergency department, women reported significantly more satisfaction with female physicians due to the increased amount of concern, trust, and overall rating of the experience [26]. These findings on specific needs of women during disasters and public health emergencies underscore the importance of ensuring adequate inclusion of women in the disaster response workforce. Given the need for gender-sensitive services in disaster response, as well as the evidence on the importance of gender-concordant service providers for women, the lack of work on the impact of gender on response readiness represents a significant gap in the literature on disaster preparedness.

Using survey data collected from a volunteer emergency management workforce organized by the Department of Veterans Affairs (VA), we tested the hypothesis that background demographic characteristics of education, age, ethnicity, and gender impacted intermediate variables of experience, attitudes about deployment, and stress. These variables were, in turn, hypothesized to predict readiness to deploy. The intermediate variables and outcome variable were constructed as multiple indicator latent variables. We were particularly interested in examining the relationship between gender and readiness to deploy in response to a disaster.

METHODS

The current literature on attitudes toward training lacks established, well-developed conceptual models [27]. Specifically, the relationship between background characteristics, organizational training, and readiness to deploy has not been well explored. Accordingly, we propose a conceptual framework that incorporates these elements.

The model builds on Bloom's "taxonomy of learning domains," which incorporates three domains of educational activities as Knowledge, Skills, and Attitude (KSA) and has been used as a tool to measure educational interventions in nursing personnel, among many others [28-30]. Francke and colleagues' conceptual model incorporates KSAs to measure the determinants for behavioral change subsequent to continuing education programs based on the interplay between background characteristics of the individual and the educational program [27]. It has been identified in nursing competencies that ongoing development of knowledge and clinical expertise helps to build skills within the KSA model [31]. In the current study, a number of questionnaire items are key indicators of the KSA domains of Bloom's model. For example, agreement with the item, "DEMPS training provided useful information about my role and responsibilities during deployment" likely reflects a level of knowledge about the requirements of deployment. Agreement with the item "I am confident in my ability to provide quality care during deployment" is an indicator for an attitude of confidence or self-efficacy that is important to successful performance in disaster response.

In our adapted conceptual model on disaster education and attitudes toward training, background characteristics and experience as a DEMPS volunteer directly influence KSAs. This, in turn, impacts readiness to deploy. Age and ethnicity were also included as background demographic variables. In addition, the proposed model includes stress as an intermediate variable, as Ejaz's conceptual model specifies background characteristics and stress as determinants of job satisfaction in direct care workers [32].

The respondent sample for this study consists of volunteers in the DEMPS program, a system of volunteers and training by which active or retired VA personnel can register for deployment for internal or external support as may be requested by federal agencies [33]. DEMPS volunteers include mental health personnel, healthcare support professionals, physicians, nurses, physician assistants, pharmacy technicians, allied health professionals, and allied support such as cooks and drivers [33].

The entire DEMPS workforce of 8,250 volunteers was invited to participate in the online survey. On October 16, 2011, an initial email invitation was sent to the DEMPS workforce and three subsequent reminder emails were sent to those who did not respond to the initial survey invitation. Each follow-up email was sent one week apart. Of those invited, 2,385 volunteers responded to the survey for a total response rate of 29%. Because of item-level nonresponse, 2079 individuals had usable data for this analysis. The study protocol was reviewed and approved by the Stanford University Institutional Review Board (IRB).

Measures

Background Sociodemographic Characteristics

Education was based on a 1–5 scale (1 = high school diploma or equivalent, 2 = Bachelor's degree, 3 = Master's degree, 4 = Doctoral degree, 5 = Medical degree). The median degree was a Master's degree and the modal response was a medical degree. *Age* was measured on a 1–6 scale (1 = 18–25, 2 = 26–30, 3 = 31–40, 4 = 41–50, 5 = 51–60, 6 = >60). The modal age range was ages 41–50. The median age range was ages 31–40. *Ethnicity* in the study was represented by minority ethnicity (21%) vs. White ethnicity (79%), coded 1 or 0 respectively. Minority ethnicity included American Indian/Alaskan native (2%), Asian or Pacific Islander (2%), African-American (11%), and Hispanic (6%). The sample was 55% female; gender was scored male = 1, female = 2. The top three occupations were "administrative, technical, or professional employee" (n = 246); Registered nurse, Level II (n = 236), and "Other administrative, technical, professional, or clerical worker" (n = 233) (not shown in table).

Intermediate Latent Variables

Experienced DEMPS Volunteer was indicated by two items: 1) How long have you been a DEMPS volunteer? (1 = less than 1 year; 2 = 1–2 years; 3 = 3–5 years, 4 = >5 years). 2) Have you ever been deployed as a DEMPS volunteer to a disaster site? (1 = no; 2 = yes).

Positive Attitude about Training was indicated by 9 items scaled 1–7 that ranged from "strongly disagree" to "strongly agree." Typical items included "The training I received was appropriate given DEMPS mission and goals" and "My DEMPS-related training events and exercises were realistic." Table 1 includes the wording for all of the items in this scale.

Table 1: Questionnaire on DEMPS training experiences

These next questions are about your DEMPS training experiences as a whole. Please select one response to indicate how much you disagree or agree with each of the following statements:							
	Strongly disagree	Disagree	Slightly disagree	Neither agree nor disagree	Slightly agree	Agree	**Strongly agree**
The training I received was appropriate given DEMPS mission and goals.	1	2	3	4	5	6	7
My DEMPS-related training events and exercises were realistic.	1	2	3	4	5	6	7
DEMPS training events and exercises were well organized.	1	2	3	4	5	6	7
DEMPS training events and exercises were good learning opportunities.	1	2	3	4	5	6	7
DEMPS training prepared me about what to expect during deployment.	1	2	3	4	5	6	7
DEMPS training provided useful information about my role and responsibilities during deployment.	1	2	3	4	5	6	7
I have been provided with sufficient training in preparation for deployment.	1	2	3	4	5	6	7
I plan to continue to volunteer with DEMPS for at least another year.	1	2	3	4	5	6	7

I would recommend volunteering with DEMPS to others.	1	2	3	4	5	6	7
All in all, I am satisfied with my training experience(s) with DEMPS.	1	2	3	4	5	6	7

Zagelbaum *et al.*

Zagelbaum *et al. BMC Emergency Medicine* 2014 14:16, doi:10.1186/1471-227X-14-16

Stress was assessed with 8 items from the Perceived Stress Scale [34]. Items are scaled from 0 (never) to 4 (very often). Typical items included: "How often have you been upset because of something that happened unexpectedly?", and "How often have you felt that you were unable to control the important things in your life?"

Outcome Latent Variable

Readiness to Deploy was indicated by 9 items scaled from 1 (strongly disagree) to 7 (strongly agree). Typical items included: "I am prepared for deployment" and "I am confident in my ability to provide quality care during deployment." Table 2 includes the wording for all of the items in this scale.

Table 2: Questionnaire on perceived readiness and preparation

These next questions are about your level of readiness and preparation for deployment. Please select one response to indicate how much you disagree or agree with each of the following statements:							
	Strongly disagree	**Disagree**	**Slightly disagree**	**Neither agree nor disagree**	**Slightly agree**	**Agree**	**Strongly agree**
I am prepared for deployment.	1	2	3	4	5	6	7
My family and/or friends support my participation as a DEMPS volunteer.	1	2	3	4	5	6	7
My supervisors and coworkers support my participation as a DEMPS volunteer.	1	2	3	4	5	6	7
I am confident in my ability to effectively respond to the deployment mobilization process.	1	2	3	4	5	6	7
I am confident in my ability to meet all administrative demands during demobilization.	1	2	3	4	5	6	7
I am confident in my ability to provide quality care during deployment.	1	2	3	4	5	6	7

I am confident in my ability to manage stress during deployment.	1	2	3	4	5	6	7
I am confident that I will know how to access mental health support if needed while deployed.	1	2	3	4	5	6	7
I feel prepared to deal with unexpected situations that may occur during deployment.	1	2	3	4	5	6	7
If an event was to occur in the next 3 months, and I am asked to deploy, I am likely to go.	1	2	3	4	5	6	7

Zagelbaum *et al.*

Zagelbaum *et al. BMC Emergency Medicine* 2014 14:16, doi:10.1186/1471-227X-14-16

Analysis

Measurement and path models were developed and tested with a latent variable covariance structure analysis using the EQS structural equations modeling program [35]. Goodness-of-fit was assessed with the Maximum-Likelihood χ^2 statistic (ML χ^2), the Comparative Fit Index (CFI), the Satorra-Bentler χ^2(S-B χ^2), the Robust Comparative Fit Index (RCFI), and the Root Mean Squared Error of Approximation (RMSEA) [36,37]. The S-B χ^2 was used in addition to the ML χ^2 because it is more appropriate and robust when the data depart from multivariate normality as the data did in this study (Mardia's normalized estimate = 139.76). The CFI and RCFI range from 0 to 1 and reflect the improvement in fit of a hypothesized model over a model of complete independence among the measured variables. Values at .95 or greater are desirable, indicating that the hypothesized model reproduces 95% or more of the covariation in the data [37]. The RMSEA is a measure of fit per degrees of freedom, controlling for sample size, and values less than .06 indicate a relatively good fit between the hypothesized model and the observed data [37].

An initial Confirmatory Factor Analysis (CFA) assessed the adequacy of the hypothesized measurement model and provided bivariate correlations among the model variables. Several items within the scales were similar, and we expected that some additional associations would be needed for fit improvement. Suggestions from the LaGrange Multiplier (LM) test for additional relationships between the error residuals within the scales were evaluated and were allowed if they made sense theoretically and logically [38]. The LM test reports supplementary modifications to the original model that will improve fit. We then tested a predictive path model that positioned the background demographic variables of education, age, minority ethnicity, and female gender as predictors of intervening variables of*Experienced DEMPS Volunteer*, *Positive Attitude about Training*, and *Stress*. In turn, all background variables predicted *Readiness to Deploy*. Correlations were allowed among the predictive background variables if they were significant. Non-significant paths and correlations in this model were trimmed gradually following the recommended procedure of MacCallum [39]. In addition to direct paths, we also examined indirect paths mediated through the intermediate latent variables.

Missing Data

About 10% of the sample was missing 1 or more individual items. To ascertain that the data were Missing Completely At Random (MCAR), the Full Information Maximum Likelihood (FIML) missing data method available in EQS that uses an EM algorithm was employed [36]. In EM imputation, parameter estimates are obtained by iterating an expectation (E) step and a maximization (M) step. FIML is the recommended data imputation method when using the EQS structural equations modeling program. Diagnostics using the models described below indicated that the missing data points were MCAR. Further, results were the same with both methods.

RESULTS

Confirmatory Factor Analysis

The final CFA had an excellent fit: maximum-likelihood solution: ML $\chi^2 = 1874.29$, 443 *df;* CFI = .96, RMSEA = .046, Robust S-B $\chi^2 = 1545.91$, 443 *df;* RCFI = .95; RMSEA = .040. All hypothesized factor loadings were significant ($p \leq .001$). Table 3 presents the factor loadings, means, and standard deviations of the measured variables. It is notable that scores on the *Positive Attitude about Training*and *Readiness to Deploy* measures were relatively high, whereas scores on *Stress* were low. Table 4reports the correlations among all of the latent variables and the demographic single-item variables in the CFA. A few correlated error residuals were added to the CFA model to improve the fit. Examining the larger correlations (all $p \leq .001$) we note that older participants reported more experience (.24), a more positive attitude about training (.10), and more Readiness to Deploy (.09). Female gender was only minimally but in some cases significantly associated with the other variables in the model (greater education, .09). Those that were experienced reported more positive attitudes (.14) and greater Readiness to Deploy (.25). Stress was associated with a less positive attitude (−.09) and less Readiness to Deploy (−.21). A positive attitude about training was highly associated with Readiness to Deploy (.47).

Table 3: Summary statistics and factor loadings of measured variables in the confirmatory factor analysis ($N = 2079$)

Measured demographic variables	**Mean (S.D). or %**	**Factor loadinga**
Education (range = 1–5)	3.01 (1.44)	NA
Age (1–6)	3.30 (0.98)	NA
Minority	21%	NA
Female	55%	NA
Latent variables		
Experienced DEMPS volunteer		
Length of time volunteer (1–4)	2.65 (0.97)	.75
Ever deployed to disaster site	18%	.64
Positive attitude about training (1–7)		
Training appropriate	5.07 (1.55)	.90
Training realistic	4.74 (1.55)	.86
Well organized	4.79 (1.50)	.87
Good learning opportunities	4.92 (1.53)	.88
Prepared what to expect	4.90 (1.59)	.90
Useful information about roles	4.94 (1.61)	.89
Training sufficient	4.55 (1.81)	.85
Recommend volunteering	5.75 (1.42)	.59
Satisfied with training	4.84 (1.81)	.89
Stress (0–4)		
Upset at unexpected	1.31 (0.70)	.66
Unable to control important things	1.18 (0.79)	.66
Felt nervous and "stressed"	1.49 (0.74)	.69
Could not cope	0.88 (0.77)	.63
Hard to control irritations	1.36 (1.05)	.42
Felt you were on top of things (R)	0.66 (0.77)	.37
Angered by lack of control	1.25 (0.75)	.70
Difficulties piling up	0.89 (0.71)	.73
Readiness to deploy (1–7)		
I am prepared for deployment	5.46 (1.67)	.79
I have family/friend support	6.08 (1.17)	.59
Support from supervisors	5.09 (1.79)	.37
Confident in my ability to respond	5.90 (1.38)	.90
Confident - administrative demands	5.89 (1.39)	.88
Confident – provide quality care	6.20 (1.16)	.81
Confident - access mental health	5.84 (1.47)	.80
Confident – deal with unexpected	5.99 (1.33)	.86

Likely to go if needed next 3 months	6.06 (1.37)	.62

[a]All factor loadings significant, p ≤ .001. NA = not applicable. R = reverse scored.

Zagelbaum *et al.*

Zagelbaum *et al. BMC Emergency Medicine* 2014 14:16, doi:10.1186/1471-227X-14-16

Table 4: Correlations among variables in model (*N* = 2079)

	1	2	3	4	5	6	7	8
1. Education	—							
2. Age	.08**	—						
3. Female gender	.09**	-.05	—					
4. Minority	-.06*	.03	-.06*	—				
5. Experienced volunteer	.00	.24**	-.01	-.05	—			
6. Positive attitude	.01	.10**	.06*	.01	.14**	—		
7. Stress	-.04	-.03	.01	.04	-.01	-.09**	—	
8. Readiness to deploy	-.02	.09**	-.06*	.02	.25**	.47**	-.21**	—

*p ≤ .05, **p ≤ .001.

Zagelbaum *et al.*

Zagelbaum *et al. BMC Emergency Medicine* 2014 14:16, doi:10.1186/1471-227X-14-16

Path Analysis

Figure 1 presents the final path model after trimming. This model has acceptable fit statistics: ML χ^2 = 1886.32, 447 *df;* CFI = .96, RMSEA = .045, Robust S-B χ^2 = 1563.41, 447 *df;* RCFI = .95; RMSEA = .039. All three intermediate latent variables (*Experienced DEMPS Volunteer, Positive Attitude about Training,* and *Stress*) significantly predicted the outcome latent variable *Readiness to Deploy*. In addition, female gender independently predicted less Readiness to Deploy ($p \leq .001$). There were also significant indirect effects on Readiness to Deploy by age ($p \le .001$), female gender ($p \le .01$), and stress ($p \le .001$). The indirect effect of female gender was positive (mediated through *Positive Attitude about Training*) although the direct effect was negative.

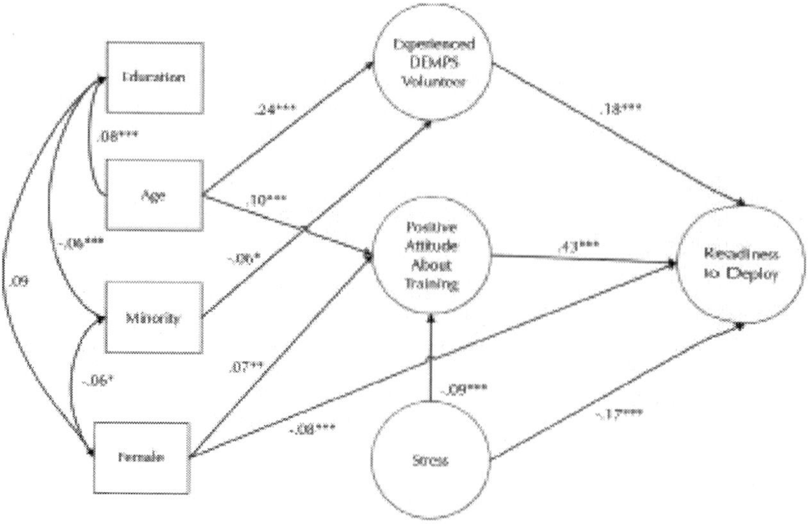

Figure 1: Model depicting significant regression paths that predict*Readiness to Deploy* among 2079 DEMPS volunteers. Large circles represent latent variables; rectangles represent single-item indicators. 1-headed arrows represent regression coefficients; 2-headed arrows represent correlations. Regression coefficients are standardized. (*= p ≤ .05, **= p ≤ .01, ***= p ≤ .001).

To assess the representativeness of the respondents to the DEMPS survey, we obtained data on demographic characteristics of the VA workforce from the 2013 VA All-Employee Survey [40]. Twenty-one percent of the respondents to the DEMPS survey were of racial/ethnic minority backgrounds, compared with 33% of the overall VA workforce. Fifty-five percent of DEMPS survey respondents were female, compared with 59% of the VA workforce. The modal age category for the DEMPS survey respondents was 41–50 years old, compared with 50–59 years old in the VA workforce.

DISCUSSION

Recruiting and retaining volunteers who are ready to deploy is essential to mounting an effective response during an emergency or disaster. Training in disaster medicine has a direct impact on a healthcare professional's ability to be useful during an emergency [1]. Training is highly desired among emergency responders, a vast majority of whom report they would benefit from additional training activities [41]. As hypothesized and in accord with our conceptual model, the current study found that age was positively associated with *Experience as a DEMPS volunteer, Positive Attitude about Training,* and *Readiness to Deploy*. This finding is also consistent with the notion that more trained and experienced healthcare workers are more likely to feel prepared to deploy [7]. Accordingly, our findings suggest that inexperienced responders may benefit from additional training in order to increase their readiness to deploy. There was also positive association between *Positive Attitude about Training* and*Readiness to Deploy*. Measuring attitudes about training is important for emergency organizations to design better training to ensure a robust emergency response workforce.

The current study found that female gender is negatively correlated with *Readiness to Deploy*. This finding is consistent with previous studies, where related factors such as familial obligations, marital status, and spousal support during deployment have been postulated to play a significant role in overall willingness to deploy for military service [42]. Given the literature suggesting that gender is an important component of cultural competence that would allow emergency responders to effectively engage with the diverse population that may be impacted by a disaster, it is important that women be included in emergency response groups [21,26]. As such, our finding that female gender has a modest negative correlation with *Readiness to Deploy* suggests that systems such as DEMPS should consider how best to address the factors that may make women less ready to deploy than men in order to ensure adequate gender representation among emergency responders. To the extent that at least some of the gender differential relates to familial obligations, marital status, and spousal support, policies that address these concerns would also benefit male responders.

The model also identified an indirect positive relationship between female gender and *Readiness to Deploy* mediated by *Positive Attitudes toward Training*. This finding is consistent with the presented conceptual model where gender influences the KSAs that will ultimately impact healthcare workers' readiness to deploy. Our study also found a positive correlation between female gender and having a*Positive Attitude about Training,* whereas those with less positive attitudes were less ready to deploy. This finding highlights the importance of providing ongoing training support to disaster response personnel to ensure the ability to provide gender-sensitive care during an emergency. A better understanding of gender differences in readiness could inform the design and implementation of emergency management training activities. This represents another key point of interest for VA and other emergency response organizations, where efforts to increase access to training and support for female responders could lead to an overall more prepared female deployment staff.

The analysis showed that individuals who reported stress were less likely to report a positive attitude about training. This has implications for the DEMPS program, as stress has been linked to impaired work performance among military personnel, which may impact outcomes in daily occupational duties as well as activities during deployment [20]. When examining gender differences in stress, we expected women to report greater stress than men, because, in general, women are more likely to report stress, experience the physical and emotional symptoms of stress, as well as undergo increased stress over the past five years [43]. However, we did not find a correlation between gender and stress. This may be due to the fact that the DEMPS volunteer workforce generally represents a self-selected group of resilient women who are highly experienced in emergency management and disaster response work.

This study has limitations. About 10% of the sample had missing data. However, statistical analysis suggested that the missing data were missing completely at random (MCAR). Therefore, the events that led to any particular item being missing appear to be independent both of the observed variables and of unobserved parameters. In addition, the data were self-reported. Thus, *Readiness to Deploy* is based on an individual's perceptions rather than objective standards. Nonetheless, it is likely that an individual's readiness is largely influenced by the degree to which he or she believes they are able to deploy at a given point in time. The categories for the ordinal age variable were not of equal width. Age was not collected as a continuous variable, so we were unable to re-categorize it into intervals of equal length. Still, ordinal variables do not need to have equally sized categories in order to be useful analytically. We included the variable on "length of time" with DEMPS based on the notion that respondents registered with DEMPS for longer time periods would likely be exposed to both more training opportunities and more deployment opportunities; however, we realize that "length of time" is only a proxy measure of these variables. The 29% response rate is somewhat low; however, since the entire DEMPS workforce was included in the survey it is possible that nonresponders have no experience being deployed. It should be noted that we do not know whether measures of willingness to deploy in this or other studies actually predict future deployment behaviors. In addition, our study did not include measures of familial obligations, marital status, or spousal support. It is possible that the observed difference in *Readiness to Deploy* between the genders could be explained by such measures, had we been able to include these items in the study. Future work should assess the predictive validity of willingness to deploy measures, as well as the mechanisms by which gender impacts *Readiness to Deploy* by including measures of these factors.

CONCLUSIONS

As previously discussed, there is a limited body of research that has explored the relationship between personnel support programs and readiness. During large scale disasters and emergencies, there is a need to identify and mobilize personnel who require the training and skills for appropriate response [33,44]. Understanding the significance of providing training to DEMPS volunteers has several simultaneous benefits. First, it provides ongoing support for volunteers to be able to provide excellent care in the event of a disaster. Second, it addresses a historical disparity in female healthcare providers who can respond to the gender-sensitive demands for a historically underserved demographic. The DEMPS volunteer program addresses an important need to identify and deploy trained health personnel in order to provide an effective response and comply with the VA's emergency management mission [45]. Our findings regarding readiness to deploy may also apply to other emergency response programs such as the

Medical Reserve Corps and the American Red Cross. Future work should explore similar relationships among these groups.

Consent

Informed consent was obtained from the respondents for the publication of this report.

AUTHORS' CONTRIBUTIONS

KH and NZ participated in the conception and design of the study and wrote the manuscript. KH/AD/JR/TN participated in the design of the study and data acquisition. JS was responsible for data analysis and wrote the methods section. RS and TN participated in the design of the study and revision of the manuscript. All authors have read and approved the final version of the manuscript.

ACKNOWLEDGEMENTS

This material is based upon work supported in part by the Department of Veterans Affairs, Veterans Health Administration, Office of Public Health. The views expressed in this article are those of the authors and do not necessarily reflect the position or policy of the Department of Veterans Affairs or the United States government.

REFERENCES

1. Merchant RM, Leigh JE, Lurie N: Health care volunteers and disaster response–first, be prepared. *N Engl J Med* 2010, 362(10):872-873.
2. Subbarao I, Lyznicki JM, Hsu EB, Gebbie KM, Markenson D, Barzansky B, Armstrong JH, Cassimatis EG, Coule PL, Dallas CE, King RV, Rubinson L, Sattin R, Swienton RE, Lillibridge S, Burkle FM, Schwartz RB, James J: A consensus-based educational

framework and competency set for the discipline of disaster medicine and public health preparedness. *Disaster Med Public Health Prep* 2008, 2(1):57-68.

3. Perry RW, Lindell MK: Preparedness for emergency response: guidelines for the emergency planning process. *Disasters* 2003, 27(4):336-350.
4. Devnani M: Factors associated with the willingness of health care personnel to work during an influenza public health emergency: an integrative review. *Prehosp Disaster Med* 2012, 27(6):551-566.
5. Institute of Medicine Committee on the Future of Emergency Care in the United States Health System: *Hospital-Based Emergency Care: At the Breaking Point*. Washington: The National Academies Press; 2007.
6. Qureshi K, Gershon RR, Sherman MF, Straub T, Gebbie E, McCollum M, Erwin MJ, Morse SS:Health care workers' ability and willingness to report to duty during catastrophic disasters. *J Urban Health* 2005, 82(3):378-388. Epub 2005 Jul 6. PubMed PMID: 16000654; PubMed Central PMCID: PMC3456052
7. Balicer RD, Omer SB, Barnett DJ, Everly GS Jr: Survey of local public health workers' perceptions toward responding to an influenza pandemic. *J Healthc Prot Manage* 2006, 22(2):1-14.
8. Irvin CB, Cindrich L, Patterson W, Southall A: Survey of hospital healthcare personnel response during a potential avian influenza pandemic: will they come to work? *Prehosp Disaster Med* 2008, 23(4):328-335.
9. Cone DC, Cummings BA: Hospital disaster staffing: if you call, will they come? *Am J Disaster Med* 2006, 1(1):28-36.
10. Loke AY, Fung WM, Liu X: Mainland China nurses' willingness to report to work in a disaster. *Am J Disaster Med* 2013, 8(4):273-282.
11. Zhiheng Z, Caixia W, Jiaji W, Huajie Y, Chao W, Wannian L: The knowledge, attitude and behavior about public health emergencies and the response capacity of primary care medical staffs of Guangdong Province, China. *BMC Health Serv Res* 2012, 12:338. Published online Sep 25, 2012. doi: 10.1186/1472-6963-12-338

12. Melnikov S, Itzhaki M: Kagan II Israeli Nurses' intention to report for work in an emergency or disaster. *J Nurs Scholarsh* 2014, 46(2):134-142.

13. Shapira Y, Marganitt B, Roziner I, Shochet T, Bar Y, Shemer J: Willingness of staff to report to their hospital duties following an unconventional missile attack: a state-wide survey. *Isr J Med Sci* 1991, 27(11/12):704-711.

14. Damery S, Wilson S, Draper H, Gratus C, Greenfield S, Ives J, Parry J, Petts J, Sorell T: Will the NHS continue to function in an influenza pandemic? a survey of healthcare workers in the West Midlands, UK. *BMC Public Health* 2009, 9:142.

15. Archer F, Seynaeve G: International guidelines and standards for education and training to reduce the consequences of events that may threaten the health status of a community – a report of an Open International WADEM Meeting, Brussels, Belgium, 29–31 October 2004. *Prehosp Disaster Med* 2007, 22(2):120-130.

16. Watt K, Tippett VC, Raven SG, Jamrozik K, Coory M, Archer F, Kelly HA: Attitudes to living and working in pandemic conditions among emergency prehospital medical care personnel. *Prehosp Disaster Med* 2010, 25(1):13-19.

17. Tippett VC, Watt K, Raven SG, Kelly HA, Coory M, Archer F, Jamrozik K: Anticipated behaviors of emergency prehospital medical care providers during an influenza pandemic. *Prehosp Disaster Med* 2010, 25(1):20-25.

18. Tucker JS, Sinclair RR, Thomas JL: The multilevel effects of occupational stressors on soldiers' well-being, organizational attachment, and readiness. *J Occup Health Psychol* 2005, 10(3):276-299.

19. Vogt DS, Pless AP, King LA, King DW: Deployment stressors, gender, and mental health outcomes among Gulf War I veterans. *J Trauma Stress* 2005, 18(2):115-127.

20. Pflanz SE, Ogle AD: Job stress, depression, work performance, and perceptions of supervisors in military personnel. *Mil Med* 2006, 171(9):861-865.

21. Richter R: Gender matters: female-specific relief efforts during disasters are key. *JEMS* 2007, 32(5):58-63.

22. Liu S, Han J, Xiao D, Ma C, Chen B: A report on the reproductive health of women after the massive 2008 Wenchuan earthquake. *Int J Gynaecol Obstet* 2010, 108(2):161-164.
23. Boscarino JA, Adams RE, Figley CR: Mental health service use after the World Trade Center disaster: utilization trends and comparative effectiveness. *J Nerv Ment Dis* 2011, 199(2):91-99.
24. Homant RJ, Kennedy DB: Police perceptions of spouse abuse: A comparison of male and female officers. *J Crim Justice* 1985, 13(1):29-47.
25. Stalans LJ, Finn MA: Gender differences in Officers' perceptions and decisions about domestic violence cases. *Women Crim Justice* 2000, 11(3):1-24.
26. Derose KP, Hays RD, McCaffrey DF, Baker DW: Does physician gender affect satisfaction of men and women visiting the emergency department? *J Gen Intern Med* 2001, 16(4):218-226.
27. Francke AL, Garssen B, Huijer Abu-Saad H: Determinants of changes in nurses' behaviour after continuing education: a literature review. *J Adv Nurs* 1995, 21(2):371-377.
28. Centers for Disease Control and Prevention Office of Public Health Preparedness and Response and Association of Schools of Public Health: *Knowledge, skills, and attitudes (KSAs) for the public health preparedness and response core competency model*. Washington: Association of Schools of Public Health; 2012.
29. O'Donnell JM, Goode JS Jr, Henker R, Kelsey S, Bircher NG, Peele P, Bradle J, Close J, Engberg R, Sutton-Tyrrell K: Effect of a simulation educational intervention on knowledge, attitude, and patient transfer skills: from the simulation laboratory to the clinical setting. *Simul Healthc* 2011, 6(2):84-93.
30. Bloom B: *Taxonomy of Educational Objectives, Handbook I: The Cognitive Domain*. New York: David McKay Co Inc.; 1956.
31. American Association of Colleges of Nursing: *Graduate-Level QSEN Competencies Knowledge, Skills and Attitudes*. Washington: American Association of Colleges of Nursing; 2012.
32. Ejaz FK, Noelker LS, Menne HL: The impact of stress and support on direct care workers' job satisfaction. *Gerontologist* 2008, 48(suppl 1):60-70.

33. Emergency Management Strategic Healthcare Group: *VHA HANDBOOK 0320.03: Disaster Emergency Medical Personnel System (DEMPS) Program and Database*. Washington: Veterans Health Administration, Department of Veterans Affairs; 2008.
34. Cohen S, Kamarck T, Mermelstein R: A global measure of perceived stress. *J Health Soc Behav* 1983, 24(4):385-396.
35. Bentler PM: *EQS 6 Structural Equations Program Manual*. Encino, CA: Multivariate Software, Inc.; 2006.
36. Satorra A, Bentler PM: Corrections to Test Statistics and Standard Errors in Covariance Structure Analysis. In *Latent Variables Analysis: Applications for Developmental Research*. Edited by Von Eye A, Clogg CC. Thousand Oaks: Sage; 1994:399-419.
37. Li-tze H, Bentler PM: Cutoff criteria for fit indexes in covariance structure analysis: Conventional criteria versus new alternatives. *Struct Equation Model Multidiscip J* 1999, 6(1):1-55.
38. Chou C-P, Bentler PM: Model modification in covariance structure modeling: A comparison among likelihood ratio, Lagrange multiplier, and Wald tests. *Multivariate Behav Res* 1990, 25(1):115-136.
39. MacCallum RC: Specification searches in covariance structure modeling. *Psychol Bull* 1986, 100:107-120.
40. Benzer JK, Young G, Stolzmann K, Osatuke K, Meterko M, Caso A, White B, Mohr DC: The relationship between organizational climate and quality of chronic disease management. *Health Serv Res* 2011, 46(3):691-711.
41. Balicer R, Omer S, Barnett D, Everly G: Local public health workers' perceptions toward responding to an influenza pandemic. *BMC Public Health* 2006, 6(1):99.
42. Kelley ML, Hock E, Jarvis MS, Smith KM, Gaffney MA, Bonney JF: Psychological adjustment of Navy mothers experiencing deployment. *Mil Psychol* 2002, 14(3):199-216.
43. American Psychological Association: *Stress in America: are teens adopting adults' stress habits?*. Washington: American Psychological Association; 2014.
44. Desforges JF, Waeckerle JF: Disaster planning and response. *N Engl J Med* 1991, 324(12):815-821.

45. Dobalian A, Callis R, Davey VJ: Evolution of the Veterans Health Administration's role in emergency management since September 11, 2001.

Chapter 3

Healthcare Worker Competencies for Disaster Training

Edbert B Hsu[1], Tamara L Thomas[2], Eric B Bass[3], Dianne Whyne[1], Gabor D Kelen[1], and Gary B Green[1]

[1]Department of Emergency Medicine, Johns Hopkins University School of Medicine, Baltimore, MD, USA

[2]Department of Emergency Medicine, Loma Linda University School of Medicine, Loma Linda, CA, USA

[3]Department of Medicine, Evidence-based Practice Center, Johns Hopkins University School of Medicine, Baltimore, MD, USA

ABSTRACT

Background

Although training and education have long been accepted as integral to disaster preparedness, many currently taught practices are neither evidence-based nor standardized. The need for effective evidence-based disaster training of healthcare staff at all levels, including the development of standards and guidelines for training in the multi-disciplinary health response to major events, has been designated by the disaster response community as a high priority. We describe the application of systematic evidence-based consensus building methods to derive educational competencies and objectives in criteria-based preparedness and response relevant to all hospital healthcare workers.

Methods

The conceptual development of cross-cutting competencies incorporated current evidence through a systematic consensus building process with the following steps: (1) review of peer-reviewed literature on relevant content areas and educational theory; (2) structured review of existing competencies, national level courses and published training objectives; (3) synthesis of new cross-cutting competencies; (4) expert panel review; (5) refinement of new competencies and; (6) development of testable terminal objectives for each competency using similar processes covering requisite knowledge, attitudes, and skills.

Results

Seven cross-cutting competencies were developed: (1) Recognize a potential critical event and implement initial actions; (2) Apply the principles of critical event management; (3) Demonstrate critical event safety principles; (4) Understand the institutional emergency operations plan; (5) Demonstrate effective critical event communications; (6) Understand the incident command system and your role in it; (7) Demonstrate the knowledge and skills needed to fulfill your role during a critical event. For each of the cross-cutting competencies, comprehensive terminal objectives are described.

Conclusions

Cross-cutting competencies and objectives developed through a systematic evidence-based consensus building approach may serve as a foundation for future hospital healthcare worker training and education in disaster preparedness and response.

BACKGROUND

Although healthcare worker training has long been accepted as an integral part of disaster preparedness, traditional training practices have not been systematically developed, rigorously examined or objectively tested. Only in recent years, under the increased scrutiny of accelerated institutional and governmental preparedness efforts, has the emerging sciences of emergency preparedness and medical education converged and a body of evidence concerning effective practices for healthcare workforce education begun to arise. The development of higher standards for workforce education based on evidence-based practices, sound educational theory and quantitative outcome measures represents an important gap to be filled in national preparedness efforts.

Providing the estimated twelve million healthcare workers in the United States with effective disaster preparedness training poses several major challenges. First, best practices to be taught must be identified and recognized. Second, specific target audiences and the content they should be taught must be defined. Instructional content should be tailored to meet the training requirements for different job categories. Effective multidisciplinary disaster response demands acquisition and application not only of factual knowledge but also complex concepts, multi-level decision making, and specific technical skills. The evaluation of whether these skill sets have been efficiently conveyed and effectively acquired presents its own inherent challenges. Finally, differences among healthcare workers such as prior training, work experience, baseline abilities and cultural background directly impact training effectiveness and must be taken into consideration for the training of large groups. The application of a systematic, competency-based approach can provide the solid foundation to build a comprehensive program of training and evaluation required to meet these challenges.

As there is currently no accepted standard for healthcare worker training in disaster response, a number of programs have adopted different formats to achieve their stated training and educational goals. Several well-recognized courses and related training materials have been aimed at improving standardization and accessibility [1-4]. To date, these practical courses fulfill an important demand but have not yet been validated or incorporated as part of standard medical training. A recent systematic literature review noted the difficulty in drawing firm conclusions about the effectiveness of specific types of hospital disaster training due to the limited numbers of studies, marked heterogeneity of training methods and weaknesses in study design and evaluation [5]. In 2001, a consensus panel of disaster medicine experts strongly recommended that education on disasters should be formalized and evidence-based and that evaluation of education and interventions must be improved [6]. Given this rapidly developing field, a number of competencies and recommendations have been made public over the past several years for clinicians, hospital workers, hospital leaders, nurses, public health workers and, health professions students but each offer considerably less detail in terms of describing measurable objectives to conduct competency-based training [7-11]. The development of professional standards and educational programs based on both the evidence and sound educational theory remains an important gap to be filled. In this paper, we propose a competency-based approach with specific measurable objectives derived by a national expert panel as a paradigm for healthcare worker disaster preparedness and response training.

METHODS

This article focuses on the development of key competencies and terminal objectives for training of all healthcare workers in disasterpreparedness. A competency was defined as a complex combination of knowledge, attitudes, and skills demonstrated by individuals that are critical to the effective and efficient function of an organization [12]. A terminal objective was defined as the demonstrable performance following the completion of a set of instruction.

The development of competencies for healthcare worker disaster response training and education incorporated an evidence-based consensus building process with the following steps: (1) review of literature on relevant content areas; (2) structured review of existing competencies, national level courses and published training objectives; (3) synthesis of new cross-cutting competencies; (4) expert panel review; (5) refinement of new competencies and; (6) development of testable terminal objectives for each competency using similar processes covering requisite knowledge, attitudes and skills.

Literature Search

A literature search updated in January 2006 identified all peer-reviewed articles related to educational competencies for healthcare workers in disasters. This consisted of identifying sources, developing a search strategy for each source and conducting and documenting the search. Electronic database searching of PubMed and the Excerpta Medica database (EMBASE) used key terms including *disaster, mass casualty, training, education, course* and *competencies*. In addition, team members conducted a hand search of the literature focusing on reference lists of relevant reviews to ensure comprehensiveness.

Structured Review of Existing Competencies and Courses

The scope of this process involved abstracting, reviewing and cataloguing existing competencies and courses in training healthcare workers for disaster preparedness. An internet search with the terms *disaster, mass casualty, MCI, chemical, biological, radiological, CBRNE, NBC, WMD, hazmat,* merged with *training, education, course* and *competencies* was conducted in January 2006 to identify existing competencies or courses for healthcare workers in disasters outside of the peer-reviewed literature. Each course or competency set was then abstracted and reviewed by the study team to determine whether it directly pertained to disasters, whether it addressed healthcare workers and whether there was an established partnership/ endorsement by a national professional organization or governmental agency. Additional relevant texts, manuals and regulatory requirements on training of hospital staff for disaster preparedness cited by these courses or published competency sets were reviewed. Courses or competency sets that did not include training for disaster response, did not target healthcare workers, or were solely sponsored by local organizations or institutions were excluded.

Synthesis of New Competencies

Current educational theory on training, curriculum and competency development was reviewed with recognized educational experts and incorporated into the development process [13-15]. Each existing competency set was reviewed for appropriateness to the target audience, defined as all healthcare workers. The core concepts from each existing competency set were identified and abstracted with grouping of common elements. A new set of cross-cutting competencies and terminal objectives were drafted and internally reviewed by the team. (Figure 1) The term, "cross-cutting", is used here to describe the competencies which may be applied to related, but distinct target audiences within the field of healthcare workers. Such groups may include, but are not limited to first receiver physicians, first receiver nurses, other first receiver staff, critical event leadership, non-first receiver physicians, nurses, technical support staff and administration.

Cross-Cutting Competencies for Healthcare Workers
1. Recognize a potential critical event and implement initial actions 2. Apply the principles of critical event management 3. Demonstrate critical event safety principles 4. Understand the institutional emergency operations plan 5. Demonstrate effective critical event communications 6. Understand the incident command system and your role in it 7. Demonstrate the knowledge and skills needed to fulfill your role during a critical event

Figure 1: Cross-cutting competencies for healthcare workers.

Expert Panel Review

The team identified 12 nationally recognized experts, which included broad representation from relevant professional organizations, hospitals, academic centers, and government agencies. Two forms were completed by the expert panel including a competency assessment form and a terminal objectives assessment form.

The competency assessment form included questions designed to address clarity, appropriateness and relative importance of each of the draft competencies. It was also designed to elicit any other possible competency areas that had not been considered. Each response was assigned a score of one (strongly agree) to five (strongly disagree). The scores in each category from the respondents were averaged. Through the process, each draft competency scoring an average of 2.5 or less in all categories was selected with minor modifications, while any scoring over an average of 2.5 in any category was to undergo revision and be resent to the expert panel for additional comment. Specific expert panel comments were reviewed, discussed and incorporated by the team into the new competencies.

The terminal objectives assessment form was designed to address the clarity and relevance of each proposed terminal objective. Each respondent was asked to complete the form, rating the clarity and relevance of a particular terminal objective for the specified target audience from one (strongly agree) to five (strongly disagree). The scores in each category from the respondents were averaged. Each draft terminal objective scoring an average of 2.5 or less for clarity was selected with minor modifications. In contrast to the cross-cutting competencies, not all terminal objectives were designed to be relevant to the full spectrum of healthcare workers. Thus, the scores for relevance were used to identify the specific terminal objectives that should be restricted to certain target audiences (ie. those that are not relevant to all healthcare workers). As with the draft competencies, specific expert panel comments on the proposed terminal objectives were reviewed, discussed and incorporated by the team.

Refinement of New Competencies

Incorporating expert panel feedback through a modified Delphi process, the team developed and refined the final competencies and terminal objectives. These competencies and terminal objectives form the framework for healthcare worker critical event preparedness training, including determination of the content, training models and teaching techniques best suited for job specific training.

RESULTS

The literature search revealed relatively few articles describing educational competencies for training of healthcare workers in disaster preparedness and response. Several described the development process and competencies for health professions students, public health workers and, nurses in emergency response [11,16-24]. However, no articles were identified that outlined disaster training competencies with specific measurable objectives. Outside of the peer-reviewed literature, the structured review of existing courses and competencies identified six sets of published competencies pertaining to the disaster training among different types of healthcare professionals, including public health workers, medical students, first receivers and registered nurses [25-30]. In spite of their importance, none included specific measurable objectives in addressing the broader spectrum of healthcare workers.

Following review and synthesis of the core concepts from the published competencies described above, seven new draft competencies were evaluated by the expert panel. No additional competency areas were suggested by the expert panel. Expert panel opinion in the form of comments and scores ranging from one (strongly agree) to five (strongly disagree) were compiled. For clarity, each of the proposed competencies averaged a score of 2.8 or less (range 1.4 – 2.8). Regarding the relevance for the defined target audience, each of the proposed competencies scored 2.1 or less (range 1.4 – 2). In terms of importance, each of the proposed competencies scored 2.5 or less (range 1.0 – 2.5). For clarity, each of the proposed terminal objectives averaged a score of 2.5 or less (range 1.2 – 2.5). The scores for relevance of the proposed terminal objectives ranged from 1.1 – 3.6.

Through this process, a total of seven cross-cutting competencies and twenty-one terminal objectives for critical event training of all healthcare workers were developed. Since disaster training must integrate multi-disciplinary training with different levels of training, a cross-cutting competency approach can serve as a valuable starting point. Cross-cutting competencies can form the basis for standardized training and unify the disciplines and skill levels involved.

Competency 1

Recognize a Potential Critical Event and Implement Initial Actions

The term, critical event, denotes any situation which threatens to disrupt the ability of an organization to maintain continuity of operations. Critical events include, but are not limited to disasters and emerging infectious diseases. An essential component of appropriate disaster response is the ability for the healthcare worker to recognize a critical event and know what to do – specifically who should be notified and how a disaster plan is activated. It is vital to provide early event recognition and early response mobilization in order to minimize the damage of the critical event. In addition, they should recognize triggers that precipitate reporting to the appropriate personnel.

TO1.1 – (Recognition)

Given scenarios that may be encountered in the course of normal professional duties, the participant should be able to identify all potential critical events and their event type. *To be successful, the participant must correctly identify all potential critical events among a list of scenarios.*

TO1.2 – (Notification)

Given a potential critical event scenario, the participant should be able to identify the appropriate authorities to be notified, recognize the appropriate notification steps and identify the key information to be reported. *To be successful, the participant must correctly identify the appropriate notification steps, information to be reported and correct reporting authority.*

TO1.3 – (Protection)

Given a description of a specific potential critical event, the participant should be able to list the immediate actions needed to protect personal, environmental and public safety. *To be successful, the participant must correctly identify standard safety precautions as well as additional precautions that may be needed for potential chemical, biological and radiological events.*

T O1.4 – (Mobilization)

Given a specific critical event scenario, the participant should be able to make recommendations for emergency response needs prior to disaster plan activation (mobilization). *To be successful, the participant must identify the specialized personnel and equipment that may be needed for the type of event and the preparation steps required for mobilization.*

TO1.5 – (Synthesis)

Given a simulated workplace scenario, the participant will apply knowledge of potential critical event recognition and immediate response needs to perform the appropriate notification, safety and mitigation actions for that event. *To be successful, the participant must correctly identify the type of potential critical event, identify the appropriate safety precautions for that event type and perform the appropriate simulated notification and mobilization actions.*

Competency 2

Apply the Principles of Critical Event Management

In order for a facility to successfully manage all critical events, healthcare workers should understand the essential elements of an effective preparation and response including the appropriate actions to be performed.

TO2.1 – (Management)

Given a list of disaster terms and management activities, the participant will be able to identify the phases of critical event management and match the activities to the appropriate phase. *To be successful, the participant must be able to correctly recognize the phases from among a list of disaster terms and match the activities to the correct phase.*

TO2.2 – (Preparedness)

Given a critical event scenario, the participant will be able to apply their knowledge of disaster preparedness to identify the key components of preparedness and recognize appropriate preparedness activities. *To be successful, the participant must be able to correctly identify the components of disaster preparedness, and select the appropriate preparedness activities for each preparedness component.*

TO2.3 – (Response)

Given a critical event scenario, the participant will be able to apply their knowledge of disaster response to identify the key components of response and recognize appropriate response activities. *To be successful, the participant must be able to correctly recognize the components of disaster response and select the appropriate activities for each response component.*

TO2.4 – (Recovery)

Given a critical event scenario, the participant will be able to apply their knowledge of disaster recovery to identify the key components of recovery and recognize appropriate recovery activities. *To be successful, the participant must be able to correctly recognize the components of disaster recovery and select the appropriate activities for each recovery component.*

Competency 3

Demonstrate Critical Event Safety Principles

A critical response component for healthcare workers is the ability to protect themselves during a disaster event who in turn will protect the facility and its resources. This encompasses elements of self and scene safety, security issues related to potentially large numbers of victims and contamination, as well as other risks or imminent threats.

TO3.1 – (Safety)

Given a critical event scenario, the participant should be able to demonstrate knowledge of critical safety principles by identifying safety threats and appropriate actions. *To be successful the participant must be able to correctly identify appropriate responses to safety threats.*

TO3.2 – (Security)

Given a critical event scenario, the participant should be able to demonstrate knowledge of security principles by identifying security threats and appropriate actions. *To be successful the participant must be able to correctly identify appropriate responses to security threats.*

Competency 4

Understand the Institutional Emergency Operations Plan

Familiarity with the institutional emergency operations plan is essential for individuals working within the institution to support and implement an effective, coordinated course of action during any critical event.

TO4.1 – (Purpose)

The participant should be able to identify the purpose and components of an EOP in critical event response. To be successful, the participant must be able to correctly identify the purposes of the EOP and its components.

TO4.2 – (Components)

Given an institutional scenario, the participant will be able to apply knowledge of critical event planning to outline an institutional EOP. *To be successful, the participant must be able to identify the necessary EOP components and functions for the scenario presented.*

Competency 5

Demonstrate Effective Critical Event Communications

Communication is a vital element to a successful critical event response. Healthcare workers need to recognize how poor communication can undermine response effectiveness and learn effective critical event communication skills.

TO 5.1 – (Communication Overview)

Given a critical event scenario, the participant should be able to apply knowledge of communications to fulfill basic communication needs including identification of appropriate timing, content, recipients, and modalities. *To be successful, the participant must correctly identify the appropriate communication steps, information to be reported, correct reporting authority, and alternative modalities.*

TO5.2 – (Communication Implementation)

Given an institutional scenario, the participant should be able apply knowledge of communications to outline a communications plan. *To be successful, the participant must outline the complex communication needs for a critical event.*

Competency 6

Understand the Incident Command System and Your Role in it

Effective critical event response requires successful integration of internal and external (local, state, and federal) participants. A recognizable and unified command and control structure is essential to achieve this.

TO6.1 – (ICS Overview)

Given a critical event scenario, the participant will be able to recognize their role in the incident command system and identify the corresponding responsibilities and limits of their authority. *To be successful, the participant should be able to identify ICS defined individual tasks and scope of responsibility.*

TO6.2 – (ICS Implementation)

Given a simulated incident command scenario, the participant should be able to apply knowledge of the incident command system to outline a command structure and interpret incoming information to make appropriate command decisions. *To be successful, the participant should be able to correctly identify lines of authority, recognize data needs and select appropriate command decisions.*

Competency 7

Demonstrate the Knowledge and Skills Needed To Fulfill Your Role during a Critical Event

Healthcare workers responding to a critical event require specific knowledge and skills. These encompass triage, personal protection, decontamination, and treatment. Injury pattern recognition and syndrome recognition that suggests the use of particular agents is essential to correct response.

TO7.1 – (Triage Skills)

Given an institutional scenario and descriptions of mock victims, the participant will be able to apply knowledge and skills concerning disaster triage systems to rapidly assign victims to appropriate triage categories. *To be successful, the participant must recognize the appropriate triage system and correctly assign triage levels to all victims within a given timeframe.*

TO7.2 – (PPE Skills)

Given a critical event scenario and descriptions of mock victims in a simulation, the participant will be able apply knowledge and skills concerning personal protective equipment (PPE) to successfully select, don, and monitor the appropriate level of PPE. *To be successful, the participant must recognize levels of PPE as well as correctly choose and simulate utilization of the appropriate level of protection.*

TO7.3 – (Decontamination Skills)

Given a critical event scenario, the participant will be able to apply knowledge and skills regarding decontamination to select, demonstrate, and monitor the correct method(s) of decontamination. *To be successful, the participant must correctly identify the decontamination level as well as select and demonstrate appropriate decontamination techniques.*

TO7.4 – (Diagnosis/Treatment)

Given a description of a patient presentation in a critical event scenario, the participant will be able to apply knowledge and skills regarding critical event diagnosis and treatment to identify critical event related syndromes and causative agents and select the appropriate treatment(s). *To be successful, the participant must correctly diagnose the presented clinical syndrome, identify the causative agent and select the appropriate treatment(s).*

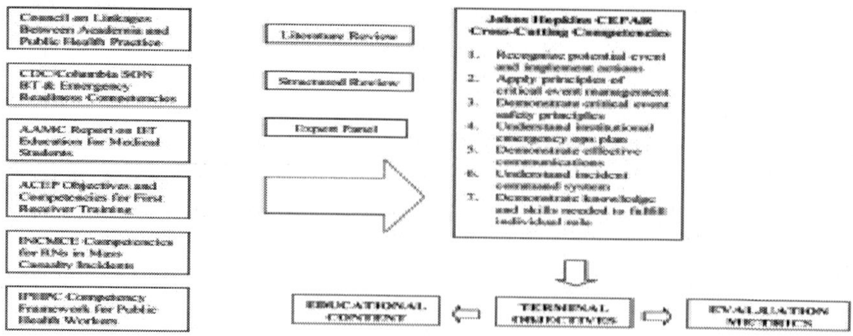

Figure 2: Development of cross-cutting competencies.

DISCUSSION

One of the highest priorities identified by the disaster response community in recent years has been to "develop standards and guidelines for education and training in the multi-disciplinary health response to major events that threaten the health status of a community" [31]. The need for rapid and effective disaster training of healthcare staff at all levels is now widely recognized and mandated by the Joint Commission on Accreditation of Healthcare Organizations (JCAHO) [32]. However, differences among healthcare workers such as cultural and educational backgrounds, prior training, work experience and abilities all contribute significant challenges to the standardization of disaster training and education. Training requirements for each job description requires appropriate modification of educational content.

Furthermore, effective disaster response demands that selected healthcare workers be equipped not only with knowledge but specific technical skills and decision-making abilities. A competency based approach provides the framework to conduct this type of flexible training.

Competency based approaches to training have been widely implemented and have gained acceptance in medical education in recent years. The advances of competency based education have been well described in the literature [33-38]. For example, the Accreditation Council for Graduate Medical Education (ACGME) has implemented core competency projects to facilitate physician training in specific knowledge and skills. Accordingly, this model has been endorsed by a number of organizations including the American Board of Medical Specialties [33]. This trend reflects an increased focus on curriculum design, establishment of measurable outcomes, adequacy of training, and program evaluation.

The application of competency based education for training healthcare workers in disaster response and other public health emergencies is a relatively new concept. Given the rapidly developing nature of the disaster field, several competency sets have recently become available since this process was conducted. For example, the Columbia School of Nursing, Center for Health Policy and the National Center for Disaster Preparedness at the Columbia University Mailman School of Public Health have described competencies for hospital workers, hospital leaders, clinicians, and nurses as part of a broader competency development program that began with the competencies for public health workers [7]. While competency statements for public health workers, medical students, and various hospital staff in emergency preparedness have been previously outlined, important differences exist in the derivation methods, degree of inclusiveness in development processes, level of detail and how the term competency is defined [25-30]. However, to our knowledge, this is the first set of cross-cutting competencies with specific measurable objectives derived using evidence-based consensus building techniques that addresses the training of all healthcare workers in disaster response. To this end, we have incorporated a systematic review of all existing related competencies and peer-reviewed literature, nationally recognized disaster courses, educational theory and expert panel consensus.

Deriving training content from a single set of cross-cutting competencies offers both theoretical and practical advantages over utilization of distinct competency sets for each type of healthcare worker being trained. First, the multidisciplinary nature of critical event preparedness and response and the importance of an integrative approach to training can not be overstated. All healthcare workers require training to develop the fundamental knowledge and skills to function independently but also to act in part of a coordinated response effort. Cross-cutting competencies unite the common underpinnings of requisite training for all healthcare workers; a departure from historically discipline-focused training.

Given the importance of problem solving and decision-making skills, the cross-cutting competencies and terminal objectives are designed to emphasize higher levels of learning such as synthesis and performance, rather than strictly knowledge acquisition [13]. In accordance with educational theory, each terminal objective was constructed so that performance may be readily tested or evaluated [13-15]. Each terminal objective can be easily mapped to the most appropriate target audience(s), with different target audiences receiving the content that will allow that group to best fulfill the relevant competency statement as applied to their job. For example, critical event leadership has different training needs from front-line physicians who in turn have different training needs from community-based practitioners. Thus, while cross-cutting competencies apply to all healthcare workers, certain terminal objectives such as those related to PPE and decontamination skills deemed essential for first receivers would be unlikely required of healthcare administrators.

Future steps will include application of this structured evidence-based process to further divide each terminal objective into enabling objectives, a series of clear statements of component knowledge, skills and/or attitudes, each with easily measurable and well-defined performance characteristics.

CONCLUSIONS

The cross-cutting competencies and accompanying objectives developed using this systematic evidence-based consensus building approach may serve as a paradigm for healthcare worker disaster training and education. While reaching similar general conclusions to other competency work that has been conducted, this framework offers the opportunity for greater standardization in training and evaluation by facilitating a uniform approach to derivation of objectives, content and evaluation.

AUTHORS' CONTRIBUTIONS

Each author has participated in the preparation of this manuscript. (EBH contributed in study concept and design, analysis and interpretation of the data, drafting of the manuscript, critical revision of the manuscript and obtaining funding, TLT contributed in study concept and design, analysis and interpretation of the data, drafting of the manuscript and critical revision of the manuscript, EBB contributed in study concept and design, analysis and interpretation of the data, and critical revision of the manuscript, DW contributed in acquisition of the data, analysis and interpretation of the data, and administrative and material support, GDK contributed in study concept and design, obtaining funding, administrative and technical support and study supervision, GBG contributed in study concept and design, analysis and interpretation of the data, drafting of the manuscript, critical revision of the manuscript, obtaining funding and study supervision.)

ACKNOWLEDGEMENTS

This work was supported by program funding from the Agency for Healthcare Research and Quality (AHRQ) Grant No. UO1 HS14337. The contents of this document are intended to provide guidance on hospital healthcare worker disaster training and should not be taken to represent standards of care. No statement in this document should be construed as an official position of the Agency for Healthcare Research and Quality or of the U.S. Department of Health and Human Services.

REFERENCES

1. United States Army Medical Research Institute of Chemical Defense: Medical Management of Chemical and Biological Casualties Course.[https://ccc.apgea.army.mil/courses/in_house/brochureMCBC.htm] Accessed on May 14, 2004
2. American Medical Association: National Disaster Life Support (NDLS). [http://www.ama-assn.org/ama/pub/category/12606.html] Accessed on May 14, 2004
3. Advanced HAZMAT Life Support (AHLS)[http://www.ahls.org] Accessed on May 14, 2004
4. Centers for Disease Control[http://www.bt.cdc.gov/agent/smallpox/training] Accessed on June 30, 2004
5. Hsu EB, Jenckes MW, Catlett CL, Robinson KA, Feuerstein CJ, Cosgrove SE, Green G, Guedelhofer OC, Bass EB: "Training of Hospital Staff to Respond to a Mass Casualty Incident" Evidence Report/ Technology Assessment Number 95. (Prepared by the JHU EPC under Contract No. 290-02-0018). *AHRQ Publication No. 04-E015- Rockville, MD: Agency for Healthcare Research and Quality*2004.
6. Seynaeve G, Archer F, Fisher J, *et al.*: Education Committee Working Group, World Association for Disaster and Emergency Medicine. International standards and guidelines on education and training for the multi-disciplinary health response to major events that threaten the health status of a community. *Prehospital Disaster Med* 2004, 19(2):S17-30.
7. Columbia School of Nursing: Competencies Projects and Incident Command. [http:/ / cpmcnet.columbia.edu/ dept/ nursing/ research/ ResCenters/ chphsr/ btcomps.html] Accessed on January 28, 2006
8. International Nursing Coalition for Mass Casualty Education: Educational Competencies for Registered Nurses Responding to Mass Casualty Incidents.[http://www.mc.vanderbilt.edu/nursing/coalitions/INCMCE] Accessed May 14, 2004
9. Core Competencies for All Public Health Workers [http:/ / www.cumc.columbia.edu/ dept/ nursing/ institutes-centers/ chphsr/ compbroch.pdf] Accessed on January 28, 2006

10. CDC Bioterrorism and Emergency Readiness: Competencies for All Public Health Workers. [http:/ / cpmcnet.columbia.edu/ dept/ nursing/ institutes-centers/ chphsr/ btcomps.pdf] Accessed on January 28, 2006
11. Markenson D, DiMaggio C, Redlener I: Preparing health professions students for terrorism, disaster, and public health emergencies: core competencies. *Acad Med* 2005, 80(6):517-26.
12. HRSA grant guidance, Nelson JC, Essien JK, Latoff JS, Wiesner PJ: *Collaborative competence in the public health agency: defining performance at the organizational and individual employee levels*. 1997.
13. Kern DE, Thomas , Howard DM, Bass : *Curriculum Development for Medical Education: A Six-Step Approach*. Baltimore: The Johns Hopkins University Press; 1998.
14. Bloom BS, Engelhart MD, Furst EJ, Hill WH, Krathwohl DR, Eds: *Taxonomy of Educational Objectives. The Classification of Educational Goals, Handbook I: Cognitive Domain*. New York: David McKay Company, Inc; 1956.
15. Competency-to-Curriculum Toolkit: Developing Curricula for Public Health Workers. Public Health Workforce Development Annual Meeting.[http://www.mailman.hs.columbia.edu/CPHP/cdc/Comptencies_tool_kit.htm] Accessed May 14, 2004
16. Morse SS: Building academic-practice partnerships: the Center for Public Health Preparedness at the Columbia University Mailman School of Public Health, before and after 9/11. *J Public Health Manag Pract* 2003, 9(5):427-32.
17. Gebbie K, Merrill J: Public health worker competencies for emergency response.*J Public Health Manag Pract* 2002, 8(3):73-81.
18. Gebbie K, Qureshi K: Emergency and Disaster Preparedness: Core Competencies for Nurses: What Every nurse should but may not know. *American Journal of Nursing* 2002, 102(1):46-51.
19. Gebbie K, Merrill , Hwang I, Gupta M, Btoush R, Wagner M: Identifying individual competency in emerging areas of practice: an applied approach. *Qual Health Res* 2002, 12(7):990-9.

20. Wisniewski R, Dennik-Champion G, Peltier JW: Emergency preparedness competencies: assessing nurses' educational needs. *J Nurs Adm* 2004, 34(10):475-80.

21. Rottman SJ, Shoaf KI, Dorian A: Development of a training curriculum for public health preparedness. *J Public Health Manag Pract* 2005, 11(6 Suppl):S128-S131.

22. Horney JA, Sollecito W, Alexander LK: Competency-based preparedness training for public health practitioners. *J Public Health Manag Pract* 2005, 11(6 Suppl):S147-S149.

23. Tachinamai T, Takemura S, Sone T, Segami K, Kato N: Competences necessary for Japanese public health center directors in responding to public health emergencies. *Nippon Koshu Eisei Zasshi* 2005, 52(11):943-56.

24. Chmar JE, Ranney RR, Guay , Haden NK, Valachovic RW: Incorporating bioterrorism training into dental education: report of ADA-ADEA terrorism and mass casualty curriculum development workshop. *J Dent Educ* 2004, 68(11):1196-9.

25. Council on Linkages Between Academia and Public Health Practice: Core Competencies for Public Health Professionals. [http://www.trainingfinder.org/competencies] Accessed May 14, 2004

26. Columbia School of Nursing Center for Health Policy: *Bioterrorism and Emergency Readiness: Competencies for All Public Health Workers. New York.* 2002.

27. Association of American Medical Colleges: *Training Future Physicians About Weapons of Mass Destruction: Report of the Expert Panel on Bioterrorism Education for Medical Students.* Washington, D.C; 2003.

28. American College of Emergency Physicians: *Final Report: Developing Objectives, Content, and Competencies for the Training of Emergency Technicians, Emergency Physicians and Emergency Nurses to Care for Casualties Resulting from Nuclear, Biological, or Chemical Incidents.* Washington, D.C.: Department of Health and Human Services, Office of Emergency Preparedness; 2001.

29. International Nursing Coalition for Mass Casualty Education: Educational Competencies for Registered Nurses Responding to

Mass Casualty Incidents.[http://www.mc.vanderbilt.edu/nursing/coalitions/INCMCE] Accessed May 14, 2004

30. Illinois Public Health Preparedness Center Competency Framework[http://www.uic.edu/sph/prepare/docs/compframe.htm] Accessed May 2004

31. Amour SJ, Bastone P, Birnbaum M, Garrett C, *et al.*: Education issues in disaster medicine: summary and action plan. *Prehospital Disaster Med* 2001, 16(1):46-9.

32. Joint Commission on Accreditation of Healthcare Organizations: Comprehensive Accreditation Manual for Hospitals: The Official Handbook. In *Oakbrook Terrace*. IL: Joint Commission Resources; 2003. EC-21

33. Accreditation Council for graduate Medical Education (ACGME): [http://www.acgme.org/outcome] *Outcome project: enhancing residency education through outcomes assessment*. 2002. Accessed May14, 2004

34. Batalden P, Leach D, Swing S, Dreyfus S: General competencies and accreditation in graduate medical education. *Health Aff (Millwood)* 2002, 21(5):103-11.

35. Carraccio C, Wolfsthal SD, Englander R, Ferentz K, Martin C: Shifting paradigms: from Flexner to competencies. *Acad Med* 2002, 77(5):316-7.

36. Leung WC: Competency-based medical training: review. *BMJ* 2002, 325(7366):693-6.

37. Long DM: Competency-based residency training: the next advance in graduate medical education. *Acad Med* 2000, 75(12):1178-83.

38. Centers for Disease Control and Prevention: Developing Objectives, Content and Competencies for the Training of Emergency Medical Technicians, Emergency Physicians, and Emergency Nurses to Care for Casualties Resulting from Nuclear, Biological, or Chemical (NBC) Incidents. [http://www.phpppo.cdc.gov/docs/library/2001]

Chapter 4

Shaping Global Agendas on Climate Risk Management and Climate Services: an IRI Perspective

James W Hansen, Stephen Zebiak, and Kevin Coffey

International Research Institute for Climate and Society, Columbia University, Palisades, NY, USA

ABSTRACT

This paper provides a historical overview of the development and mainstreaming of global agendas on climate risk management and climate services from the vantage point of the International Research Institute for Climate and Society (IRI), and present two examples that illustrate how efforts to mainstream these agendas shaped both the institute and the broader communities that it sought to engage. In the first example, we trace developments that led to the emergence of a global agenda on climate services, including creation of the Climate Services Partnership (CSP). The CSP is an informal, open network of interested climate information users, providers, donors and researchers, serves as a platform for knowledge sharing and collaboration to advance climate service capabilities worldwide. The second example is mainstreaming climate risk management within the international agricultural research-for-development community. The CGIAR research program on Climate Change, Agriculture and Food Security (CCAFS) was structured to include a research theme on "Adaptation through Managing Climate Risk," which put climate risk management on the same footing within the agenda as adaptation to future climate change. We conclude with several lessons drawn from IRI's involvement in the processes that shaped global agendas around climate risk management and climate services.

BACKGROUND

Over the past 15 years, the focus and strategy of the International Research Institute for Climate and Society (IRI) evolved – from climate prediction and applications, toward a more holistic approach to climate risk management and to climate services – in parallel with the agenda of the broader climate research and applications community that it has sought to engage (Goddard et al.2014). IRI perspectives on climate risk management and on climate services grew from the institutional experience of working toward its mission "to enhance society's capability to understand, anticipate and manage the impacts of climate in order to improve human welfare and the environment". We highlight here a few key points about the IRI's strategy for research, service and training, which are further detailed and illustrated in the companion papers of this journal issue.

The initial focus of IRI was on advancing seasonal climate prediction and its application. From the beginning, IRI work engaged both research and practical application always in the context of real decision or policy settings. Experience working at the interface between research and practice quickly revealed that practitioner communities rarely found climate information products developed by climate researchers and operational agencies (including IRI) to be understandable or useable. Some form of translation inevitably was needed to cast information into a form that addressed the right questions and informed practical decision-making processes. Innovation in the decision process was often also needed in order to take advantage of new information and knowledge about climate. Additionally, policy or institutional constraints were often found to hinder implementation of new climate-informed practices. The opportunity to overcome these challenges required substantial investment in engagement among researchers, information providers and practitioners that, when effective, enabled the "co-discovery" of approaches and supporting information to effect climate-informed practice and policy. We found that the most effective way to start the collaboration process was to focus on understanding the problems and decisions that new services would address. This learning served both as inspiration and motivation for IRI articulating its work and its objectives in a different way. Through on-the-ground experience, the institutional view on our own work and its potential value shifted from a focus on "applications" of climate forecasts and information, toward enabling better decisions and policies where climate-related risks and opportunities could be effectively assessed and managed. IRI developed and began to promote its concept and approach to climate risk management (CRM). Moreover, IRI reformulated its mission and institutional objectives, and even its name (from "International Research Institute for Climate Prediction" to "International Research Institute for Climate and Society"), in the context of CRM.

As IRI gained experience, it increased its investment in key institutional partnerships and engagement in international processes relating to climate-related challenges. These partnerships provided fertile ground for dialogue, debate and further learning that shaped the interest and perspectives that IRI has brought to the more recent international discourse and activities around climate services.

IRI's participation in the development of global agendas on climate services and climate risk management, which significantly shaped its own agenda, gives the institution a unique historic perspective. This paper provides a historical overview of the development and mainstreaming of global agendas on climate risk management (focused on agriculture and food security) and climate services from the vantage point of the IRI (summarized in Figure 1), and illustrates how IRI's approach to research and partnership influenced both the institute and the broader communities that it sought to engage. Section 2 traces IRI's contribution in the global discourse on climate services and to key events that led to the UN Global Framework for Climate Services, and the launch of a global Climate Services Partnership. Section 3 describes IRI's role in the development and leadership of the CGIAR research program on Climate Change, Agriculture and Food Security, which has contributed to mainstreaming climate risk management within the international agricultural research community. The paper concludes with some lessons learned through participating in agenda mainstreaming processes.

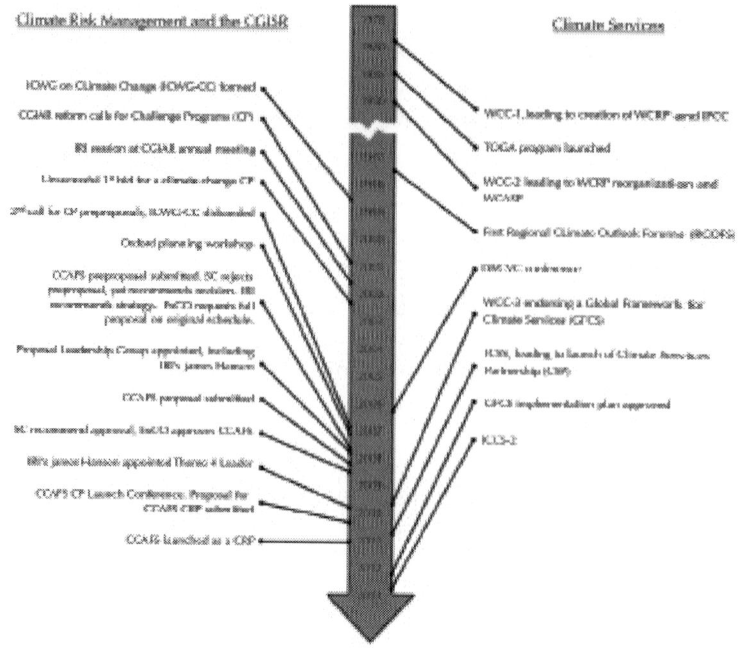

Figure 1: Time line of significant events that shaped the development global agendas for climate risk management for agriculture, and climate services.

RALLYING A GLOBAL COMMUNITY AROUND CLIMATE SERVICES

A growing appreciation for the importance of climate variations to societies, the need to better manage the associated impacts and potential opportunities, and the relevance of this for development objectives led to the growing interest and action around climate services. The Climate Services Partnership (CSP) draws on this growing interest and engagement across a global community. The CSP – an outcome of the first International Conference on Climate Services (Columbia University, New York, October 2011) – is an informal, non-governmental platform for promoting and advancing climate services throughout the world. Made up of researchers, providers, users and funders of climate services, the CSP aims to support climate services capabilities through strengthening knowledge capture and exchange, fostering collaboration and supporting institutional linkages across relevant agencies and programs. This initiative provides a platform for connecting initiatives and actors within the larger climate services agenda that is taking shape today, into a growing community of practice.

The concept and vision for climate services have evolved considerably over recent years, but rest on climate programs and activities over many decades (Zillman 2009). Here we review some aspects of this history, with attention to how and why IRI became increasingly involved.

The Roots of Climate Services

Recent international dialogue and actions around climate services build on a history of climate activities over several decades. Many national meteorological services included climate offices dating as early as the 1950s. For the most part these services focused on the provision, analysis and occasionally interpretation of historical climatological data in the context of particular "applications" such as agriculture and transportation. Climate research advanced significantly in the 1970s, with the development and increased use of global climate models that provided a new means to study climate processes, but also enabled researchers to address questions of climate sensitivity and climate change particularly in response to increasing greenhouse gas concentrations.

Building on this growing body of research, the first World Climate Conference (WCC-1) was convened in 1979. The findings stated, "Having regard to the all-pervading influence of climate on human society and on many fields of human activity and endeavor, the Conference finds that it is now urgently necessary for the nations of the world: (a) To take full advantage of man's present knowledge of climate; (b) To take steps to improve significantly that knowledge; (c) To foresee and prevent potential man-made changes in climate that might be adverse to the well-being of humanity" (Proceedings of the World Climate Conference. Geneva 1979). A principal outcome was the creation of the World Climate Programme (WCP), comprised of four component programs that addressed climate research, data, impacts and applications. Its aim was to promote and coordinate efforts to better observe, model and understand the climate system and its relation to human wellbeing. The WCP subsequently supported the establishment of the Intergovernmental Panel on Climate Change (IPCC), charged with ongoing scientific assessment of climate change, its social and economic impacts and potential response strategies.

The dimensions of climate research, and eventually climate services, expanded significantly in the 1980s, as work on understanding, modeling and predicting aspects of the El Niño/Southern Oscillation (ENSO) phenomenon intensified. Under the auspices of the World Climate Research Program, the Tropical Ocean – Global Atmosphere (TOGA) program was launched in 1985. Work under the TOGA program was highly influential, demonstrating not only a theoretical basis for predictability of climate on seasonal-to-interannual time scales associated with ENSO, but also predictions of this phenomenon validated retrospectively, and in real time (Anderson et al. 1998). By 1995, TOGA left in place an operational observing system consisting of moored buoys, island tide gauges, surface drifters, a volunteer ship observing program and a variety of satellite observations, all to support continuing routine seasonal predictions. The TOGA period represented a new era for the climate community, leading to a much more ambitious climate agenda. Following closely on the TOGA program, the concept of an international institution dedicated to both the improvement of seasonal prediction capability and its practical use toward societal benefit culminated in the founding of IRI (then the International Research Institute for Climate Prediction).

The beginnings of this expanded climate agenda were already in place as the second World Climate Conference (WCC-2) was convened in 1990. WCC-2 resulted in a restructuring of the World Climate Programme, introducing the explicit dimension of climate services by formulating the World Climate Applications and Services Program (WCASP), with a remit for operational climate information, prediction and analysis systems, including user liaison. WCC-2 reformulated the WCP program on climate impacts to include consideration of impacts and responses. Both moves broadened the emphasis from the production to the use of the information.

Under the new WCASP, initiatives were introduced that focused on the delivery of climate forecasts and information to user communities. The Climate Information and Prediction Services (CLIPS) project was aimed at developing capacity for delivering climate information for improved decision-making, focusing primarily on development of training activities and curricula. Regional Climate Outlook Forums were introduced in 1997 (a major El Niño year with strong climate-related impacts in many regions), and were designed to bring together research, operational and "user" communities around a process to produce a consensus regional climate outlook and consider associated socio-economic impacts and response strategies (Basher et al. 2001). IRI was invited into, and became very active in, both of these initiatives as an institution that had pioneered seasonal climate and impact prediction methodologies, products and tools at the global and regional scale[a] (Barnston and Tippett 2014).

[a]The probabilistic forecast product format that IRI first introduced became the standard format for most of the consensus outlooks of the RCOFs.

More services and applications activities were envisioned under the World Climate Program, but the process failed to deliver adequate resources and support throughout the decade and a half following WCC-2. The World Climate Research Program was, however, active throughout the period. Under its Climate Variability and Predictability (CLIVAR) program, the Working Group on Seasonal to Interannual Prediction was established in 1999. IRI scientists have served in the role of Chair or member of WGSIP since the beginning, as well as in other CLIVAR roles[b].

[b]For example, L. Goddard is currently co-Chair of the CLIVAR Scientific Steering Group.

Building Momentum around Climate Services

LWCVC Conference

After more than a decade hiatus following the second World Climate Conference, the concept emerged for an international interdisciplinary conference on decision processes in climate applications. IRI participated in the early discussions and planning for this conference, voicing strong support for using the opportunity to highlight emerging experience in climate risk management and user engagement processes[c]. Eventually, IRI proposed and was accepted as a co-sponsor of the conference, with WMO and the Finnish Meteorological Service.

[c]This was made possible through a formal relationship between IRI and WMO, established the preceding year. Building on informal relations over several years, the WMO and IRI signed a Memorandum of Understanding in 2004 expressing intention *"to collaborate in the delivery of effective climate services, including research in providing climate information and products, development and evaluation of appropriate decision tools, impacts analysis, communication, institutional and policy analysis, demonstration projects involving stakeholders and end-users, and related capacity-building and education activities"*.

The conference, "Living with Climate Variability and Change: Understanding the Uncertainties and Managing the Risks" (LWCVC) was convened in 2006 to "review the opportunities and constraints in integrating climate risks and uncertainties into decision-making in the core socio-economic sectors" (WMO (World Meteorological Organization) 2009a). It emphasized bringing user community perspectives together with climate scientist perspectives, drawing participation from multiple sector groups (agriculture, water, public health, energy, disasters) as well as experts in decision-making. This, together with the focus on decision processes and decision contexts distinguished LWCVC clearly from previous climate conferences. IRI was able to contribute to LWCVC in several ways. Beyond IRI serving as a co-sponsor and co-organizer, IRI scientists were invited presenters or co-chairs of many of the thematic sessions.

LWCVC provided an opportunity for presentation, discussion and debate among a diverse but invested and interested community of researchers and practitioners. From this process a number of significant conclusions emerged. The findings highlighted, for the first time, key requirements for establishing climate services that enable effective climate risk management practices. The workshop recommended that climate services should be: (a) driven by the needs and requirements expressed by relevant decision sectors, (b) developed within real-world decision contexts, (c) enabled through facilitating institutions and policies, (d) based on environmental, sectorial and socioeconomic data, (e) based on tailored climate information, (f) supported by local capacity, (g) included in planning strategies that incorporate incentives, and (h) supported by sector-specific services from National Meteorological and Hydrological Services and related institutions (WMO (World Meteorological Organization) 2009a). These ideas have permeated all subsequent discourse on climate services.

World Climate Conference-3 and the Global Framework for Climate Services

The World Climate Conference-3 (WCC-3), convened by the World Meteorological Organization in 2009, was the landmark event on the pathway toward current international initiatives in climate services. The vision statement, "WCC-3 will establish an international framework to guide the development of climate services linking science-based climate predictions and information with climate risk management and adaptation to climate variability and change throughout the world," framed climate services in the context of decision-making. This was a significant conceptual step – perhaps leap – toward the prevailing view of services that encompass decision makers as well as researchers and information providers.

An International Organizing Committee (IOC) was given the task of developing the conference scientific and policy agendas, conducting outreach to key partner organizations and sponsors, and developing appropriate communications.

Through discussion, debate and stakeholder consultation, the IOC's subcommittee for the High Level Segment defined the vision and action agenda that became the conceptual underpinnings for the Global Framework for Climate Services (GFCS). The Concept Note that was delivered to WCC-3 participants presented a rationale, objectives and a proposed programmatic structure for the GFCS that clearly built upon existing infrastructure and activities of the World Climate Programme, but also transcended them through inclusion of a mechanism to actively engage the climate services practitioner communities – the User Interface Platform[d]. IRI was invited to serve on the IOC, and through this mechanism contributed to the planning of WCC-3 and the visioning for GFCS. IRI interventions sought to broaden the User Interface Platform concept to include fostering institutional partnerships, cross-disciplinary research, innovation, decision-support tools, knowledge capture, evaluation, establishment of good practices, education and capacity building. This advice was based on IRI's own experience and that of other organizations already working in the "boundary space" of climate services development.

[d]The Framework Structure included 4 components: Observations and Monitoring; Research, Modeling and Prediction; Climate Services Information System; and the User Interface Program (WMO (World Meteorological Organization) 2009a).

WCC-3 drew more than 2500 participants, including numerous heads of state, UN agencies and international organizations (with significant IRI involvement[e]). A wide-ranging Expert Segment addressed research findings on sector-based needs and applications, climate science and information, climate risk management and climate adaptation, and various societal perspectives on climate services. The key outcome of the conference was the WCC-3 Declaration, which endorsed the GFCS as proposed through the IOC, and set in motion a process creating a High Level Task Force[f] to undertake consultations and make recommendations for next steps in implementation (WMO 2009).

[e]Associated with WCC-3, the book "Climate Sense" was published (WMO (World Meteorological Organization) 2009b). In this volume 3 papers were published with IRI lead authorship.

[f]Draft Terms of Reference and suggested membership for the Task Force were prepared in advance by the IOC.

Thus, the international climate agenda was transformed and energized around a new vision for climate services that is still in play today. Progress continues to further define and implement the GFCS, and to establish appropriate governance and mobilize resources to support it. While much of the current implementation activity rests within intergovernmental processes, there are continuing opportunities for nongovernmental partners such as IRI to contribute[g].

[g]IRI scientist S. Mason served on the drafting group for the Task Force report and GFCS Implementation Plan. IRI also contributed case studies, consulted with partner organizations and provided review and recommendations on the GFCS implementation.

Climate Services Partnership

Even as international momentum around climate services has been building, a range of climate services programs and institutions have been introduced at national and regional levels in response to growing awareness and demand across sectors (including agriculture, food security, water, health, energy and disaster management). Sector-based organizations and development institutions have invested in building climate information and services capacities into their programs. At the same time, research organizations have been working on developing new knowledge to support new and better climate services. Because these various initiatives have, for the most part, been pursued independently, the experience and knowledge gained has largely resided within each program.

This is the backdrop from which the idea of the CSP emerged. The initial, informal discussions among a few organizations were built on the concept of creating a space for exchange of information and experiences among active climate services stakeholder groups across sectors and regions. An international conference – the first International Conference on Climate Services (ICCS) – was proposed as the means to start this process. IRI worked with an expanded organizing committee to develop ideas for the conference and its outcomes. The intent was for the conference to be action-oriented, engaging organizations that were active and in climate services in some manner, with genuine interest in sharing their experience and learning from others. From early in the planning process, the notion of some sort of continuing process was held as a means to foster continued exchange and collaboration expected to emerge from ICCS.

The Conference; held at Columbia University, New York, October 2011; brought together more than 100 participants, representing meteorological or climate service centers, UN agencies, humanitarian organizations, development institutions, national agencies, non-governmental organizations and universities. The Conference produced lively debate and discussion, and achieved consensus on the proposition that first motivated the conference: that establishing a means to effectively share knowledge and build collaboration would enable the community to accelerate learning, develop new capacities and establish good practices that would benefit all.

In the ICCS Conference Statement, the participants agreed to establish the Climate Services Partnership as a platform to pursue precisely this agenda. By consensus, the Climate Services Partnership was proposed as an informal and open process in order to support broad participation, and to enable the CSP to be nimble and responsive to members' interests and ideas. A Coordinating Group was established to provide review and recommendations on activities, consider membership, resourcing questions and event planning. In support of the initiative, IRI offered to form a Secretariat[h] to coordinate communications, develop knowledge resources and organize collaborative projects of the Partnership.

[h]Initially with resources provided by IRI and the Earth Institute at Columbia University. Subsequently additional support has been provided through USAID, the German Climate Services Center, the CGIAR CCAFS program, among others. With endorsement of the Partnership, IRI continues to host the Secretariat.

The initial CSP activities have included: (a) establishing a mechanism for sharing knowledge and lessons learned, and developing a set of case studies contributed by the Partnership members; (b) initiating collaborative working group activities to expand knowledge in priority areas, including economic valuation of climate services and good practices in climate services development; and (c) identifying mechanisms to support the Global Framework for Climate Services and other relevant international processes. A variety of information resources on climate services activities, programs, institutional arrangements, outputs and experiences have been developed through CSP activities.[i] Ongoing collaborative work on the economic value of climate services includes assessment of the relevant literature and the development of new resources on valuation methodologies appropriate for climate services. CSP maintains active contact with the GFCS Project Office, and has facilitated a process to provide feedback and recommendations on proposed aspects of GFCS implementation.

'An interactive database of information about climate services activities was created, based on surveys distributed through the membership. CSP, in partnership with GFCS, developed a set of case studies, authored by members, that provides more information on outputs and lessons learned. Two climate services programs have also been assessed in greater depth, in terms of cost and benefit, gaps and opportunities. These information resources can be found on the CSP website http://www.climate-services.org.

Following the recommendation of a continuing series of ICCS conferences, CSP convened the second International Conference on Climate Services (ICCS2) in September 2012, together with local host (and CSP member), Climate Services Center, Germany. ICCS2 brought together a larger and more diverse attendance than did its predecessor (including 200 participants from 40 countries, and greater representation of the private sector and user communities). It provided opportunity for review, discussion and deliberation on initial CSP activities; and proposed additional CSP activities including: knowledge capture on climate services user perspective, institutional and business models; establishing good practices and minimum standards in science, user engagement and evaluation of climate services; capacity development; engaging NGO programs and resources; linking financial services with climate services; and principles for public-private partnerships in climate services.

CSP plans to continue to convene an annual international conference in the future, as well as support other relevant thematic or regional meetings. There has also been encouragement for developing additional, low-cost means for discussion, debate and information sharing. Toward this end, CSP has established a web-based facility – the CSP Knowledge Exchange - providing an alternative, accessible means for presenting and discussing programs, concept papers, and opinions relating to climate services.

The CSP represents a novel, grassroots initiative to advance the climate services enterprise from research, to practice, to impact.

Although CSP has established new resources, gained financial support, and engaged considerable participation, there remains great deal more that this initiative could accomplish. Its informal, nongovernmental structure makes it easy to engage academia, civil society, private sector partners and governmental organizations. Based on the interest and participation of its members, and resources they can attract, CSP has the potential to create a truly rich repository of climate services knowledge and information, a clearinghouse for continually-evolving good practices, and a dynamic collaboration space for new ideas and innovations. If it is successful, it will not only serve as a preferred resource for its individual members, but will also contribute to advancing the scientific underpinnings, implementation, assessment and improvement of climate services globally. With due attention to how such work could most effectively link to, and support, formal climate services development, especially the Global Framework for Climate Services, these CSP contributions can assume even greater significance and impact.

The CSP offers IRI a means to build upon and amplify its own work, engage new partnerships, and contribute to building global capacity in climate services and climate risk management in new ways. IRI sees its own mission and objectives reflected in CSP. It is squarely in the interest of IRI to support and contribute to CSP as a primary mechanism for contributing to its international climate services agenda.

SHAPING A GLOBAL AGENDA AROUND CRM FOR AGRICULTURE

A CRM Approach to Adapting Agriculture to a Changing Climate

Within agriculture, a framework and methodology for analyzing risk and its implications for decision-making in the agriculture sector were largely in place by the late 1970s (Anderson et al.1977). Yet climate risk management (CRM) gained identity as a concept that crosses sectors only in the last decade, with IRI prominent among the institutions that framed and promoted the concept. A few common features characterize the IRI approach to CRM (Goddard et al. 2014; Hansen et al. 2007; Hellmuth et al. 2007; Baethgen 2010). It recognizes climatic uncertainty as a fundamental challenge; and seeks to quantify and, where possible, reduce the uncertainty that decision-makers face. Managing climate risk often involves a portfolio of interventions including, for example, climate-informed technologies that reduce vulnerability to climate variability, and climate-informed policy and market-based interventions that transfer risk from vulnerable populations. Climate information – including historical observations, monitoring of current conditions and prediction at the range of relevant lead times – plays a prominent role in IRI's work on CRM. IRI's approach to CRM focuses on the full range of variability, seeking to protect against the impacts of adverse extremes, while also capitalizing on opportunities in years or seasons when climatic conditions are favorable.

Bridging Adaptation and Development

Long (30-100 year) climate change scenarios and impact studies that dominated the first three IPCC assessments may have been useful to frame political discourse toward mitigation policies, but are not relevant to the planning horizons of vulnerable farmers, or of governments of poorer countries grappling with immediate climate-sensitive development challenges. Few climate-sensitive agricultural decisions have planning horizons longer than about two decades, and most are much more immediate. Early policy under the UNFCCC compounded this dilemma by defining adaptation narrowly in terms of the anthropogenic change component of climate, and restricting the main adaptation funds to the verifiable additional adaptation costs imposed by anthropogenic climate change, while the cost of development that would target the sources of climate vulnerability had to be borne elsewhere (Pielke 2005; Khan and Roberts 2013; Ayers and Huq2009). This "additionality" policy, which was meant to ensure that industrialized countries provide adaptation funds beyond existing development assistance commitments, was seen as an obstacle to poor countries securing funds for promising adaptation options that also addressed immediate climate-related development challenges. For adaptation interventions with longer planning horizons, the considerable uncertainty inherent in projections of local changes in climate (particularly precipitation) complicates decision-making, and the challenge of justifying requests for adaptation funds. By focusing on immediate actions that target vulnerability and build resilience to the impacts of climate across time scales, CRM is increasingly recognized as a crucial bridge between immediate development challenges and longer-term adaptation.

Adaptation through a Resilience Lens

Applying a CRM approach to adapting agriculture to a changing climate raises several challenging issues. Consider the case of a stressed smallholder farming system in a risk-prone environment. The same factors – a weak asset base, and dependency on a narrow range of weather-dependent sources of livelihood– tend to make the household vulnerable to both future climate change and current climate shocks. The strong likelihood that climate change will increase the frequency or intensify of damaging extremes, such as droughts, floods or heat waves, increases the urgency of transformation toward a livelihood strategy that is less weather-dependent, more stable and more prosperous. Yet in these settings, extreme events are prone to reverse the gains from agricultural development investments by eroding farmers' productive assets. Evidence suggests that poverty traps often lock poorer farmers, in high-risk environments, into highly vulnerable livelihood strategies. The poverty traps literature highlights the existence of threshold levels of assets that act as tipping points between chronic poverty and the potential for asset accumulation (Barrett and Carter 2006). Climate variability and resulting risk appears to contribute to poverty traps in several distinct ways, including acting as a barrier to accessing available production technologies and market opportunities (Hansen et al. 2011).

The agricultural literature on risk emphasizes the risk perceptions and preferences of decision-makers, and how they interact with probability distributions of variables such as production and income in response to stochastic drivers (Anderson et al. 1977; Hardaker et al. 2004). While this framework is useful, it is not sufficient for dealing with the implications of risk for the climate change adaptation challenge, and particularly how the dynamic interactions between risk, decisions, and household wealth or assets, evolve over time. The possibility that these dynamic interactions are affected by critical thresholds or "tipping points" presents a particular challenge. For these reasons, climate risk management in the context of adaptation to progressive climate change is increasingly expressed in terms of resilience, which offers a conceptual and analytical framework for dealing with the evolution of nonlinear systems characterized by critical thresholds, subjected to stochastic variability and uncertain trends.

Resilience from an ecological perspective was originally defined as "the persistence of relationships within a system and is a measure of the ability of these systems to absorb changes of state variables, driving variables, and parameters, and still persist" (Holling 1973). When applied to human systems, the concept of resilience had to be extended to include both decision processes that allow deliberate and anticipatory reconfigurations of the system (i.e., adaptation), and the very real possibility that the "original state" is undesirable (Engle 2011; Bahadura et al. 2013).

A more recent framing of resilience from a development perspective provides a theoretical link between near-term risk and climate change impacts for smallholder farmers (Barrett and Constas2013). In the context of poverty traps, the authors discuss resilience in term of a livelihood trajectory that is subject to stochastic risk (e.g., from climate variability) that inhibits asset building for smallholder farmers. For rainfed farming systems vulnerable to seasonal variability, climate shocks can cause repetitive cycles of asset loss that threaten progress out of poverty traps, leaving households in a chronic state of food insecurity. In this case, a resilient rural household does not necessarily return to its initial state following a shock. Rather, a resilient household is one that is moving toward a more positive state, accumulating enough assets to escape chronic poverty, while protected from setback by shocks. This interpretation moves beyond resilience as the ability of a system to retain its current form regardless of climate variability (Berman et al.2012). Instead, resilience to climate shocks can allow farmers to build wealth, which facilitates a transformation towards livelihoods that do not require asset depleting coping strategies.

Thinking about the adaptation challenge through a resilience lens has several practical implications for efforts to support smallholder agriculture. It provides a framework for thinking about what combination of development and safety net interventions should be considered in a particular context, to move communities onto a positive livelihood trajectory that will be well-adapted to future climatic conditions, while protecting them from falling back in the face of climate shocks. If a farming system is characterized by a poverty trap, adaptation efforts must either increase farmers' assets above the poverty trap threshold, or change the conditions that led to the poverty trap. Understanding how climate-related risk undermines agricultural development and contributes to poverty traps may suggest promising new interventions to improve the success of development efforts. In the presence of poverty traps, different forms of safety nets may be needed to protect the poor and vulnerable from destitution or abandonment, and to protect the non-poor from falling into chronic poverty.

A partnership between the IRI and the international agricultural research community provided an opportunity to develop and apply the IRI's evolving thinking about CRM in the context of a new applied research program on Climate Change, Agriculture and Food Security (CCAFS).

The CCAFS Story

CCAFS – a program of the CGIAR in partnership with the global change research community – is the world's largest research effort to address the challenges that a changing climate imposes on agriculture and food security across the developing world. The CGIAR[j] is a network of 15 International Agricultural Research Centers (IARCS), and the funding and governance bodies that support them. CGIAR research played a major role (Evenson and Gollin 2003) in the Green Revolution, which led to near tripling of global food production between 1961 and 2000 when global population doubled[k], and dramatically reduced poverty and hunger particularly in Asia and Latin America. The aggregate return on investment in CGIAR research, in terms of development impact, has been high throughout its history (Raitzer and Kelley 2008; Renkow and Byerlee 2010). The CGIAR evolved significantly since it was established with three IARCS in 1971. As the number of IARCS increased, its mission broadened from increasing global food availability to include first rural poverty reduction and then natural resource sustainability.

[j]Prior to its recent restructuring, "CGIAR" was an acronym for Consultative Group for International Agricultural Research.

[j]Based on FAOSTAT data (http://faostat3.fao.org/faostat-gateway/go/to/home/E).

[k]The CGIAR defines an "outcome" as a change in behavior (e.g., change in policy, investment, programs or practice) of a stakeholder that results from a research output, and that that is expected to contribute toward impacting some measure of societal wellbeing or environmental quality.

Since 2000, IRI has viewed the CGIAR as the most strategic partner for advancing climate risk management for agriculture. The opportunity to partner with the CGIAR in the development of CCAFS was a breakthrough for the IRI's efforts to advance CRM for agriculture.

The Challenge Program Phase

A set of reforms initiated in 2001 within the CGIAR (CGIAR 2000) set the stage for CCAFS. These included the introduction of Challenge Programs as a new model for collaborative research that sought to strengthen synergies and collective action among the IARCs; and to expand the CGIAR's research agenda, partnerships and funding beyond the core work of the IARCs. Challenge Programs were designed as "time-bound, independently governed programs of high-impact research that target CGIAR goals on complex issues of overwhelming global or regional significance that require partnerships among a wide range of institutions to deliver their products". The Challenge Program goal of moving beyond the traditional mandates of IARCS, on emerging issues of global concern, was quite relevant in the case of climate change. While climate is not explicit in the mandate of any of the IARCs, attention to climate pervades much of the CGIAR's research because agriculture is so weather-dependent. The CGIAR had already formed an Inter-Center Working Group on Climate Change (ICWGCC) in 1998, as a mechanism to address the growing issues of climate change adaptation and mitigation. An initial call for proposals led to the creation of four Challenge Programs; but a proposed Climate Change Challenge Program, led by the ICWGCC, was not approved.

CCAFS came out of a second call for Challenge Program pre-proposals in 2006. The CGIAR Alliance, representing directors of the IARCs, agreed to dissolve the ICWGCC and pursue a Challenge Program, in partnership with the global change research community represented at the time by the Earth Systems Science Partnership (ESSP). The process started with a small meeting (Oxford, UK, Feb 2007) of representatives from the CGIAR and global change research community that proposed a strategy and preliminary thematic structure. The pre-proposal that came out of the process was not accepted. Yet the CGIAR Executive Council considered the topic sufficiently important to request a full proposal, on a tight original time frame. IRI responded quickly to the resulting challenge with a set of recommendations for moving the process forward, which were well received and eventually adopted. As a result, IRI was invited to participate in a 9-person Leadership Group, representing the CGIAR and the global change research community, tasked with developing the CCAFS proposal.

Objectives of the proposed program (CCAFS 2009) targeted: (a) critical knowledge gaps about how to enhancing food security, livelihood and environmental goals; (b) adaptation options; and (c) supporting decision-making – in response to a changing climate. A research agenda to achieve these objectives was structured within six Themes (Table 1). The Leadership Group concluded that two separate adaptation themes were justified – focused on managing current climate risk, and on adapting to progressive change – to ensure sufficient leadership to cover the range of research challenges, options for intervention, and required partnerships. Analysis and stakeholder consultation led to selection of three focus regions: Eastern and West Africa, and South Asia. Other features of the proposed program included a partnership between the CGIAR and ESSP (representing the global change research community at-large), a focus on food systems in addition to agricultural production, and governance that maintained independence from the IARCs.

Table 1: Research theme in the CRP and challenge program phases of CCAFS

CGIAR research program phase		**Challenge program phase**	
1	Adaptation to progressive climate change	5	Adaptation pathways under progressive climate change
2	Adaptation through managing climate risk	4	Adaptation pathways based on managing current climate risk
3	Pro-poor climate change mitigation	6	Poverty alleviation through climate change mitigation

4	Integration for decision making	1	Diagnosing vulnerability and analyzing opportunities
		2	Unlocking the potential of macro-level policies
		3	Enhancing engagement and communication for decision-making

Hansen et al.

Hansen et al. Earth Perspectives 2014 1:13, doi:10.1186/2194-6434-1-13

A competitive process led to IRI hosting the leadership of Theme 4: Adaptation pathways based on managing climate risk". Work began late 2009, and CCAFS was formally launched at a conference in Nairobi in May 2010.

The CRP Phase

During the period between submission of the Challenge Program proposal and the official launch of the program, the CGIAR began a major restructuring (Secretariat, CGIAR 2009). The most significant change was the creation of CGIAR Research Programs (CRPs) that would become the mechanism for organizing, reporting and funding research. The CRPs would replace all Challenge Programs. The CCAFS leadership was tasked with formulating a CRP to deal with climate change. Climate change was one of two fast-tracked CRP proposals that led the reform, and set many of the rules and norms that the remaining CRPs would follow.

The resulting CRP (CCAFS 2011; Vermeulen et al. 2011) built heavily on the CCAFS Challenge Program. The six Themes of the Challenge Program were consolidated into four (Table 1). The new CRP incorporated the new climate risk management agenda and other major areas of research, intervention and partnerships from the CCAFS Challenge Program; but also incorporated ongoing climate adaptation and mitigation work within the IARCs. Since all CGIAR research was to be mapped onto CRPs – most of which hadn't been yet been defined – some initial ambiguity about the boundaries had to be resolved. For example, while much of the CGIAR's work on developing genetics and production technology for stressed environments contributes toward climate adaptation, there was a consensus that this would generally fit better within other CRPs; CCAFS would emphasize targeting and evaluating of portfolios of production technology in the context of climate variability and change. Other significant changes to CCAFS included aspects of governance, source of funding (from individual donors to primarily CGIAR funding), and a significant increase in total funding to cover the large amount of ongoing research among the IARCS that was integrated into the program. The responsibility of Theme Leaders expanded to include reporting on relevant work across the IARCs, and ensuring that the work aligned with the CCAFS agenda and produced outcomes.[1] Although CCAFS was dissolved as a Challenge Program, the branding and leadership of CCAFS continued through the transition to CRP with only minor adjustments. This allowed strategic partnerships and activities initiated in the Challenge Program phase to continue.

Mainstreaming an Expanded Climate Risk Management Agenda

The transition of CCAFS from a Challenge Program to a CRP contributed significantly to mainstreaming climate risk management within the CGIAR. This is because the purpose of a Challenge Program was to expand the agenda and partnerships of the CGIAR, while CRPs were mainstreamed as the new mechanism organizing and resourcing research across the CGIAR. As a Challenge Program, CCAFS was explicitly designed to complement rather than replicate the core work of the CGIAR's IARCs by targeting: (a) emerging adaptation interventions that were not yet mainstreamed due to newness or knowledge gaps; (b) interventions requiring expanded upstream and downstream partnerships beyond the CGIAR's traditional partners; (c) analytical approaches to support targeting and evaluation of adaptation options developed by CGIAR centers; and (d) integrated, cross-scale adaptation approaches that would depend on the coordination, integration and economy of scale that a Challenge Program could provide (CCAFS 2009). The new CCAFS agenda, which incorporated elements of IRI's approach to CRM, became the starting point for an ambitious agenda that was mainstreamed across the IARCS. Within this new agenda, synergies between IRI's climate expertise and CRM experience, and

CGIAR's broad base of agricultural expertise and action on the ground across the developing world, opened opportunities for new areas of research, such as climate information and advisory services for smallholder farmers, bringing improved crop production forecasting into food security early warning, and an expanding range of climate-informed food security safety nets.

Expanding on one example, several CGIAR centers had researched the use of seasonal forecasts for farmers prior to CCAFS, but IRI played a key role in bringing these experiences together and adding new technical and applied knowledge. IRI experience enabled the work to incorporate forecasts that were downscaled and tailored to the needs of smallholder farmers, going beyond the information that was routinely available in target countries. Pilot projects with smallholder communities in Kenya and Senegal provided insights about how to design and communicate climate information for farmers. A conference in late 2012, that brought together practitioners from Africa and Asia, catalyzed efforts to develop and mobilize resources for regional efforts to scale up climate services for farmers (Tall et al. 2013). More recent efforts have invested in the capacity of several African meteorological institutions to provide climate information at a scale that is relevant to smallholder farmers. This line of research has connected the CGIAR with the major climate services initiatives described in the case study in Section 2, with CCAFS playing a prominent role in the Climate Services Partnership, and in the first national implementation project, targeting Tanzania and Malawi, in the Global Framework for Climate Services.

CONCLUSIONS

The extent to which IRI sought to bridge disciplines, communities and agendas – internally and externally – was groundbreaking within the climate research community. IRI's efforts to connect climate science and decision-making led to significant changes in its own agenda, mission, and even the name of the institution. In time, the same commitment to integrative partnerships and co-learning opened doors to bring some of our internal lessons into global dialogs and shape global agendas. IRI's evolving agenda had particular influence on global agendas in the areas of climate risk management and, more recently, climate services.

The IRI's experience in participating in the development and mainstreaming of new global agendas suggests a few lessons for other institutions working internationally at the interface between science and society:

First, engaging in global dialogs and partnering with larger communities can multiply impact relative to what is possible through the direct efforts of an institution's own staff. Some of the greatest advances toward its mission have resulted when the IRI played a relatively minor role in service to a larger community. Through CCAFS, for example, leveraging the considerable human capacity and reach of the CGIAR has enabled IRI to advance its work on CRM in an area where its own resources are very limited.

Second, long-term, persistent investment in relationships with strategic partners is necessary to set the stage to influence on global agendas. In both examples presented in this paper, IRI invested in strategic relationships over several years (Vaughan et al., 2014), generally long before clear avenues for significant impact were apparent. This required effort to identify where interests overlapped, and effort to add value to the work of partners. Those relationships proved to be at least as important as intellectual leadership on the issues that the IRI sought to influence.

Third, breakthroughs are likely to come through serendipity, and require awareness and responsiveness to opportunity. More often than not, IRI has been able to participate effectively when it recognized unanticipated opportunities and had the flexibility to respond quickly, but only when key relationships had already been built over time. In the case of CCAFS, a timely and strategic suggestion by IRI staff opened the door for a longer-term role in program design and implementation. The impact that the CGIAR reform process and transition of CCAFS to a CRP had on mainstreaming an expanded climate risk management agenda across the CGIAR is another example of a breakthrough through serendipity.

Fourth, the greatest advances in these global agendas were associated with new connections between communities that previously had little interaction. By expanding the CGIAR's interactions with the climate community and with the food security information and response community, CCAFS opened the door to the CGIAR to new avenues for impact. The CSP now enables sustained interaction between several major development organizations and funders, and the climate research community, leading to new opportunities and better-informed investments in climate services for development. The research agenda that CCAFS defined for the CGIAR could only be implemented through new partnerships with the climate research, climate services, and food security information response communities – partnerships that the IRI sought to enable.

One example that illustrates several of these lessons is a partnership with the Red Cross, jointly framed around "Early Warning, Early Action" (Coughlan de Perez and Mason 2014). Through several years of working together, we not only developed a refined sense of what kinds of information and services could usefully address operational needs of the disaster management community; but also came to mutual understanding of the need, and some common approaches to stakeholder engagement in climate services. This enabled both institutions to advocate for these ideas, with a common voice and greater impact, in the global dialogue and planning processes that shaped the Global Framework for Climate Services.

Finally, participating effectively in the development of global agendas requires openness to change within an institution's own agenda and culture. Before the IRI could engage effectively in international discourse, its own staff had to learn how to work effectively across disciplines, and step out of their comfort zones. As IRI interacted with an expanding and evolving community of partners, its own institutional thinking about how science interacts with other segments of society evolved significantly – away from a top-down "expert" – "user" model, and toward co-production of knowledge and co-ownership of communication processes.

AUTHORS' CONTRIBUTIONS

JH developed material on climate risk management for agriculture, framed concluding lessons, and incorporated input from other authors. SZ developed material on the climate services agenda. KC contributed to climate risk management in the context of the adaptation and resilience literature.

ACKNOWLEDGEMENTS

The authors thank the editor Dr. Gerten, guest editor Lisa Goddard, Mea Halperin and two anonymous reviewers for their constructive feedback and suggestions. The initiatives and partnerships described in this paper were made possible by NOAA's long-term investment in IRI. This publication and the work it describes were supported in part by grant/cooperative agreement NA10OAR4310210 from the U.S. National Oceanic and Atmospheric Administration (NOAA); by the CGIAR Research Program on Climate Change, Agriculture and Food Security (CCAFS), a strategic partnership of CGIAR and Future Earth; by the European Union (EU) with technical support from the International Fund for Agricultural Development (IFAD); and by the U.S. Agency for International Development (USAID). The views expressed in this publication cannot be taken to reflect the official opinions of NOAA, CGIAR or Future Earth, or USAID.

Responsible editor: Lisa Goddard

REFERENCES

1. Anderson JR, Dillon JL, Hardaker JB (1977) Agricultural decision analysis. Ames, IA: Iowa State University Press.
2. Anderson DLT, Sarachik ES, Webster PJ, Rothstein LM (1998) The TOGA Decade. J Geophys Res 103, C7:14167-14510
3. Ayers JM, Huq S (2009) Supporting Adaptation to Climate Change: What Role for Official Development Assistance? Dev Policy Rev 27:675-692 Publisher Full Text
4. Baethgen W (2010) Climate risk management for adaptation to climate variability and change. Crop Sci 50:S70-S76
5. Bahadura AV, Ibrahimb M, Tannera T (2013) Characterising resilience: unpacking the concept for tackling climate change and development. Clim Dev 5(1):55-65
6. Barnston T, Tippett M (2014) Climate Information, Outlooks, and Understanding - Where Do We Stand? Earth Perspect 1:20
7. Barrett CB, Carter MR (2006) The economics of poverty traps and persistent poverty: An asset-based approach. J Dev Stud 42(2):178-199 Publisher Full Text

8. Barrett CB, Constas M (2013) Toward A Theory of Resilience for International Development Applications. http://dyson.cornell.edu/faculty_sites/cbb2/Papers/Barrett-Constas-Oct-2013.pdf

9. Basher R, Clark C, Dilley M, Harrison M (eds) (2001) Coping with Climate: A Way Forward. Summary and Proposals for Action Palisades, New York: International Research Institute for Climate Prediction.

10. Berman R, Quinn C, Paavola J (2012) The role of institutions in the transformation of coping capacity to sustainable adaptive capacity. Environ Dev 2:86-100

11. CCAFS (2009) Climate Change, Agriculture and Food Security. A CGIAR Challenge Program. Rome and Paris: The Alliance of the CGIAR Centres and ESSP.

12. CCAFS (2011) CCAFS Program Plan. Copenhagen, Denmark: CGIAR Research Program on Climate Change, Agriculture and Food Security (CCAFS). [http://hdl.handle.net/10568/10227]

13. CGIAR (2000) Charting the CGIAR's future reshaping the CGIAR's organization. Summary of proceedings and decisions. CGIAR 2000 International Centers Week ICW2000. Washington, DC, 23-27 Oct 2000.[http://www.worldbank.org/html/cgiar/publications/icw00/icw00sop.pdf

14. Coughlan de Perez E, Mason SJ (2014) Climate Information for Humanitarian Agencies: Some Basic Principles. Earth Perspect 1:11 BioMed Central Full Text

15. Engle NL (2011) Adaptive capacity and its assessment. Glob Environ Chang 21(2):647-656

16. Evenson RE, Gollin D (2003) Crop Variety Improvement and Its Effects on Productivity. Wallingford, UK: CABI Publishing.

17. Goddard L, Bhojwani H, Robertson AW, Baethgen WE (2014) The International Research Institute for Climate & Society: Why, What and How. Earth Perspect 1:10

18. Hansen JW, Baethgen W, Osgood D, Ceccato P, Ngugi RK (2007) Innovations in climate risk management: protecting and building rural livelihoods in a variable and changing climate. J Semi-Arid Trop Agric Res 4(1): [http://www.icrisat.org/Journal/SpecialProject/sp13.pdf]

19. Hansen JW, Zebiak S, Baethgen W, Ines AVM, Osgood D (2011) Climate Risk, Information and Market Participation for African Farmers. In: Proceedings of an international conference, Nairobi, Kenya, 13-15 May 2009. Nairobi: AGRA and ILRI. pp 157-175
20. Hardaker JB, Huirne RBM, Anderson JR, Lien G (2004) Coping with Risk in Agriculture. CAB International: Wallingford.
21. Hellmuth ME, Moorhead A, Thomson MC, Williams J (eds) (2007) Climate Risk Management in Africa: Learning from Practice Palisades, New York: International Research Institute for Climate and Society.
22. Holling CS (1973) Resilience and stability of ecological systems. Ann Rev Ecol Systemat 4:1-23
23. Khan MR, Roberts JT (2013) Adaptation and international climate policy. WIREs Clim Change 4:171-189
24. Pielke RA Jr (2005) Misdefining "climate change": consequences for science and action. Environ Sci Policy 8:548-561
25. Proceedings of the World Climate Conference. Geneva (1979) WMO-No. 537, Geneva.
26. Raitzer DA, Kelley TG (2008) Benefit–cost meta-analysis of investment in the international agricultural research centers of the CGIAR. Agr Syst 96:108-123
27. Renkow M, Byerlee D (2010) The impacts of CGIAR research: A review of recent evidence. Food Policy 35:391-402
28. Secretariat, CGIAR (2009) Voices for Change: the New CGIAR. Washington DC: CGIAR Fund Office. [http://hdl.handle.net/10947/5035]
29. Tall A, Jay A, Hansen JW (2013) Scaling Up Climate Services for Farmers in Africa and South Asia Workshop Report. Copenhagen: CGIAR Research Program on Climate Change, Agriculture and Food Security (CCAFS). [http://hdl.handle.net/10568/27833]
30. Vaughan LF, Furlow J, Higgins W, Nierenberg C, Pulwarty R (2014) US Investments in International Climate Research and Applications: Reflections on Contributions to Interdisciplinary Climate Science and Services, Development, and Adaptation. Earth Perspect 1:23
31. Vermeulen S, Zougmore R, Wollenberg E, Thornton P, Nelson G, Kristjanson P, Kinyangi J, Jarvis A, Hansen J, Challinor A,

Campbell B, Aggarwal P (2011) Climate change, agriculture and food security: A global partnership to link research and action for low-income agricultural producers and consumers. Curr Opin Environ Sustain 4(1):128-133

32. WMO (2009) World Climate Conference-3: Towards a Global Framework for Climate Services. WMO Bull 58(3):162-164
33. WMO (World Meteorological Organization) (2009) Report of the WMO Conference on Living with Climate Variability and Change: Understanding the Uncertainties and Managing the Risks.
34. WMO (World Meteorological Organization) (2009) Climate Sense. Leicester, UK: Tudor Rose.
35. Zillman JW (2009) A history of climate activities. World Meteorological Organization (WMO) Bulletin 58(3):141

Chapter 5

Evacuation Planning in the Auckland Volcanic Field, New Zealand: A Spatio-temporal Approach for Emergency Management and Transportation Network Decisions

Erik Tomsen[1,4], Jan M Lindsay[1], Mark Gahegan[2], Thomas M Wilson[3], and Daniel M Blake[3]

[1]School of Environment, The University of Auckland, Private Bag 92019, Auckland, New Zealand

[2]Centre for eResearch & Department of Computer Science, University of Auckland, Private Bag 92019, Auckland, New Zealand

[3]Centre for Risk, Resilience and Renewal, Department of Geological

Sciences, University of Canterbury, Private Bag 4800, Christchurch, New Zealand

[4]Present address: CMR 464, APO, AE 09226, USA

ABSTRACT

Auckland is the largest city in New Zealand (pop. 1.5 million) and is situated atop an active monogenetic volcanic field. When volcanic activity next occurs, the most effective means of protecting the people who reside and work in the region will be to evacuate the danger zone prior to the eruption. This study investigates the evacuation demand throughout the Auckland Volcanic Field and the capacity of the transportation network to fulfil such a demand. Diurnal movements of the population are assessed and, due to the seemingly random pattern of eruptions in the past, a non-specific approach is adopted to determine spatial vulnerabilities at a micro-scale (neighbourhoods). We achieve this through the calculation of population-, household- and car-to-exit capacity ratios. Following an analysis of transportation hub functionality and the susceptibility of motorway bridges to a new eruption, modelling using dynamic route and traffic assignment was undertaken to determine various evacuation attributes at a macro-scale and forecast total network clearance times. Evacuation demand was found to be highly correlated to diurnal population movements and neighbourhood boundary types, a trend that was also evident in the evacuation capacity ratio results. Elevated population to evacuation capacity ratios occur during the day in and around the central city, and at night in many of the outlying suburbs. Low-mobility populations generally have better than average access to public transportation. Macro-scale vulnerability was far more contingent upon the destination of evacuees, with favourable results for evacuation within the region as opposed to outside the region. Clearance times for intra-regional evacuation ranged from one to nine hours, whereas those for inter-regional evacuation were found to be so high, that the results were unrealistic. Therefore, we conclude that, from a mobility standpoint, there is considerable merit to intra-regional evacuation.

INTRODUCTION

Evacuations are, and most likely will continue to be, the most common and efficient emergency management strategy when a hazardous event threatens and puts at risk the safety of those within the area (Moriarty et al. 2007). Evacuations are also becoming increasingly frequent worldwide as humans continue to develop in hazardous areas and improved technology in many countries allows for prior warnings and the movement of people before a disaster strikes (Sparks 2003, Woo and Grossi 2009). However, evacuations can produce long-term negative effects such as psychological trauma, and disruption of community cohesion and employment and economic continuity (e.g. Mileti et al. 1991, Lindell and Perry 1992, Cola 1996, Tobin and Whiteford 2002, Perry and Lindell 2003). Poorly managed evacuations tend to lead to a strong resentment of government which, in turn, decreases the ability of emergency management organisations to act effectively in the future (MCDEM2008). Therefore, effective planning of an evacuation is essential.

Volcanic eruptions are capable of producing a spectrum of hazards which are harmful to humans. These hazards range from highly destructive phenomena such as pyroclastic density currents, debris avalanches, lava flows and lahars that typically destroy everything in their path, to less destructive yet highly disruptive phenomena such as ash fall, volcanic tremor and gas release. As many volcanoes and volcanic regions around the world are already heavily populated, the most effective means of risk reduction will be to identify the most hazardous areas and evacuate the population from the danger zone prior to an eruption (Marzocchi and Woo 2007, Lindsay et al. 2011, Sandri et al. 2012).

The city of Auckland, New Zealand, with a population of ~1.5 million as of December 2012, is built within the 360 km^2 potentially active basaltic Auckland Volcanic Field (AVF) (Figure 1), with the last eruption occurring just 550 years ago (Needham et al. 2011). Auckland's geography poses significant constraints for evacuation planning. The city is located on an isthmus bounded by the Waitemata Harbour to the northeast and Manukau Harbour to the southwest. As a result, all land-based transport into and out of the city is constricted through narrow stretches of land serviced by four motorway bridges which form critical links in Auckland's transportation network (Figure 1). To date, there has been no major modelling-based study conducted on the mass evacuation capacity of Auckland. Previous evacuation planning has been mainly strategic and lacking in geospatial analysis and physical evacuation procedures that can be used operationally (Auckland CDEM Group 2008a, Auckland CDEM Group 2008b, Tomsen 2010).

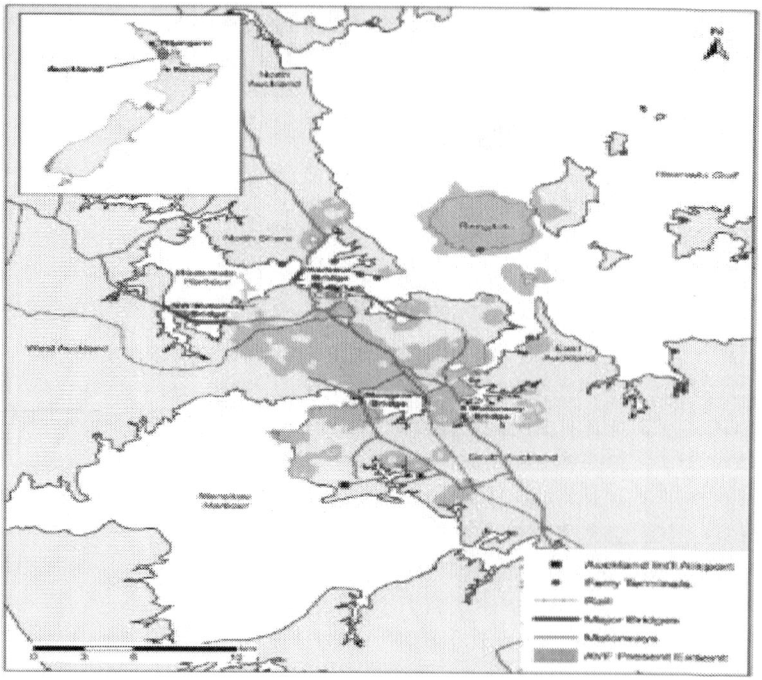

Figure 1: Map of Auckland (inset shows the location of Auckland, Hamilton and Whangarei within New Zealand), showing volcanic deposits of the Auckland Volcanic Field (representing present extent), and key infrastructure elements.

In this study we adopt a novel, non-specific approach (Shulman 2008) in considering the spatial and temporal distribution of population and transport networks across Auckland and how they affect mass evacuation planning. Spatial network analysis is used to determine the geographic functionality of major transport origin and destination points and we determine the relative vulnerabilities of the key motorway bridges to new AVF eruptions. We then assess micro-evacuation vulnerability by combining spatial network analysis with population evacuation demand to calculate evacuation capacity ratios for individuals, households and vehicles. Finally, we employ modelling using dynamic route and traffic assignment to measure evacuation attributes at a macro-scale and forecast total network clearance times. This quantitative study thus serves to fill the informational void and provides emergency management officials with a more holistic understanding of the local variations in susceptibility to mass evacuations, particularly those related to volcanic activity in the AVF.

Evacuation Planning

There is a broad body of literature on effective evacuation planning. Many studies have attempted to classify evacuations into various types (e.g. Baker 1991, Ketteridge et al. 1996, Wolshon et al. 2001, Marrero et al. 2010) and others have focussed on emergency response activities (e.g. Cova 1999, Cutter 2003, Marzocchi and Woo 2007, Moriarty et al. 2007, Shaluf 2008). The core components to this evacuation planning can be summarised as: 1) conditions under which an evacuation may be necessary; 2) 'at risk' people/communities who may require evacuation; 3) evacuation routes and destinations; and 4) the resources and time required to evacuate 'at risk' people/communities (MCDEM, 2008).

Identifying when an Evacuation is Necessary

Evacuation can be classified as an "organised, phased, and supervised withdrawal, dispersal, or removal of civilians from dangerous or potentially dangerous areas, and includes their reception and care in safe areas" (U.S. Department of Transportation 2006, p.2-1). Evacuation becomes necessary when the benefits of leaving significantly outweigh the risk of other options, such as 'sheltering-in-place'. In a volcanic context, evacuation is a response strategy – an effort to preserve human life (Marzocchi and Woo 2007, Auckland CDEM 2013). In order to assist with evacuations, plans are created in advance, identifying key personnel, areas at risk, and mitigation measures to enact (Moriarty et al. 2007). In New Zealand, the Mass Evacuation Plan (MCDEM 2008) is the key sub-national level plan which aims to detail a range of considerations and actions for the mass evacuation of people from a hazardous environment to a relative place of safety (Auckland CDEM Group 2008b). The Auckland Volcanic Field Contingency Plan is more specific and includes planning arrangements for evacuations resulting from an eruption within the AVF (Auckland CDEM 2013). According to the plan, an evacuation will be called by the Auckland CDEM Group if hazard assessment indicates urban or strategic areas may lie within 5 kilometres of the inferred eruption centre and/or there is a potential risk to life.

'At Risk' People and Communities

There is varied focus in the literature about which group or groups tend to be the most 'at risk'. Low-income populations are studied in detail by some (e.g. Morrow 1999, Chakraborty et al. 2005), while others (e.g. Bascetta 2006, Dosa et al. 2007) focus on the elderly and disabled. The *low-mobility population* (i.e. those without access to a private vehicle), however, are discussed by many and we examine this group further as they will require public modes of transport (Leonard 1985, Hushon et al. 1989, Wolshon et al. 2001). Ideally, people within an evacuation zone evacuate and people resident outside the zone shelter in place. However, evacuations are typically far more complex, with some choosing to remain within a zone, and others outside the evacuation zone voluntarily evacuating (termed *shadow evacuation,* Baker 1991).

Uncertainty as to who will stay and go (the *population evacuation demand*) makes it difficult to establish credible time estimates for those evacuating, although this is fundamental for evacuation planning. The population evacuation demand is dependent on numerous variables including external conditions such as weather, location of the hazard source and time, as well as human behavioural characteristics inherent in the population (Wolshon 2006, Tomsen 2010). Research on evacuation response rates for hurricane-based evacuations in the United States found that evacuation rates ranged between 33-97% during the same hurricane, with an average of 47.5% (Baker 1991). People in high-risk areas, on average, were found to be more than twice as likely to evacuate when compared to low-risk areas. This was attributed to two factors: people residing in high-risk areas are aware of the hazardousness of their location and/or public officials go to greater lengths to evacuate the residents of these areas (Baker 1991). Less data is available for non-compliance to shelter-in-place orders, a factor that often causes emergency management officials the most difficulty. Two recent surveys regarding such *unofficial evacuees* reported nearly 60% of respondents leaving before evacuation orders were given during Hurricanes Lili and Katrina (Lindell et al. 2005, Lindell and Prater2006).

Evacuation Routes and Destinations

Evacuation route choice is a complex decision-making process. Some researchers believe that in emergency situations, evacuees will take any possible egress route (Moriarty et al. 2007). However, others contend that people will take the most familiar routes (predominantly motorways), which often become overloaded while capacity on alternative routes remains unused (Prater et al. 2000, Dow and Cutter 2002). During Hurricane Katrina, drivers were more influenced by familiarity with the route than traffic conditions they experienced en route (Lindell and Prater 2006). As stated in the Mass Evacuation Plan for New Zealand (MCDEM 2008, p.56), "the planning process should decide upon primary and secondary evacuation routes from an anticipated affected area", and "evacuation routes should be designed with due consideration to local area hazard maps to ensure that selected routes are appropriate for anticipated hazards". Any potential bottlenecks in traffic movement should also be identified (MCDEM 2008). Many studies in the U.S. have shown that, despite the enormous demand during hurricane evacuations, many roads carry flows well below the predicted maximums (Wolshon2008). However, contraflow systems are frequently used for evacuations in the U.S. and plans are often well engineered and publicised (Wolshon 2002). Indeed, although studies on hurricane-based evacuations provide valuable information, many differences in characteristics such as risk perception, familiarity and cultural geography, mean that the results cannot be easily extrapolated to other hazards or locations (Marrero et al. 2010). Transportation modelling can be used to help with specific planning objectives, and allow the testing of various assumptions and alternatives.

During evacuations people tend to favour temporary re-location in second homes, hotel/motel accommodation or with family and friends, rather than seeking public shelter (Quarantelli 1985). However, in a mass evacuation, many of the low mobility population and those without social networks or financial resources will require assistance with accommodation from emergency management authorities. For smaller events, 'all-in-one welfare facilities' may be all that are required to service evacuees. However, when the volume of evacuees is likely to be large, separate evacuation and recovery centres may need to be established (MCDEM 2008).

Evacuation Resources and Time

The ability of a community to respond to a disaster and cope with its consequences largely depends on its level of preparedness. However, the impact on an evacuated community is reduced when evacuation is carried out in a well-managed and organised manner. During a mass evacuation, transportation networks are the most critical components of a region's infrastructure network, as they facilitate the mobility of the human population. In developed countries, private vehicles have often been the predominant form of mass evacuation (Quarantelli 1980, Drabek 1986, Lindell and Perry1992, Tierney et al. 2001, Cole and Blumenthal 2004). This is likely due to their prominence in today's society, the flexibility of route and destination choice they allow, as well as their asset value, which many evacuees seek to retain. A survey conducted in 2008 by the New Zealand Ministry of Civil Defence and Emergency Management with regard to evacuation behaviour in Auckland, confirmed this tendency. Of the 2,050 people in the survey, 91.3% would choose to leave with their own vehicle if required to evacuate due to an AVF eruption (Horrocks 2008b). Alternative forms of transportation such as trains and buses can also be used for evacuation purposes and are particularly beneficial to the low mobility population who may strongly rely on their provision.

When considering the time and resources required for evacuation, it is important to acknowledge the regular diurnal population shift which occurs in most developed countries when people travel to places of work and learning during the day and return home again at night. A national telephone survey conducted by Klepeis et al. (2001) across the U.S. demonstrated that while more than 90% of people are at home and indoors between the hours of 11 pm and 5 am, less than 35% are there from 10 am to 3 pm. We expect similar trends to occur in New Zealand, particularly in city environments, although there is little data for comparison at present. In addition to the standard diurnal shift, other spatio-temporal movement patterns exist in urban areas. On weekends and during school holidays, when many residents leave for recreational activities and travel, the population in the urban area sinks compared to its weekday highs. At other times, such as during major concerts, sporting events and conventions, the urban population may grow substantially. When evacuation time estimates are available, emergency management officials can determine how far in advance evacuation orders should be issued. This allows authorities to balance the competing demands of enduring public safety and unnecessary costs associated with imprecise or unnecessary evacuations, i.e. false alarms. Because running evacuation drills is difficult due to the large areas and populations involved, computer simulations based on various traffic analysis models offer the next best option (Franzese and Liu 2008). Current emergency management planning in Auckland assumes that a major evacuation (such as for an impending volcanic eruption) would require 48 hours for authorities to implement (this includes a pre-evacuation-call planning period). This was illustrated in the lead up to the simulated evacuation during a major 2008 exercise based on an Auckland Volcanic Field eruption, *Exercise Ruaumoko*, when civil authorities wanted to know when the 48 hour 'time window' before outbreak had been entered (Lindsay et al. 2010).

The Auckland Volcanic Field

When considering evacuation planning for a volcanic field eruption the following factors related to the hazard must be considered: the likelihood; the number of vents expected; the location(s) of the new vent; the area impacted by volcanic hazards (hazard footprint), which is dependent on the style and size of eruptive activity produced during vent opening; and how much warning will be provided by volcano monitoring systems. The two most important factors are the hazard footprint and the location of the eruption, which together allow determination of the necessary spatial extent of the evacuation zone. This section reviews the past known eruptive history of the AVF with particular focus on these two factors.

The AVF (Figure 1) is a geologically young, generally monogenetic, intraplate volcanic field made up of over 50 small basaltic volcanoes, which has been active for 250,000 years with the last eruption ~550 years ago (Lindsay et al. 2010, Needham et al. 2011, Shane et al.

2013). Being generally monogenetic in nature, each vent is typically only active for a single eruption sequence and new eruptions usually occur in a different location from those before. To date there have been no spatio-temporal trends identified for vents in the AVF. Recent algorithmic analysis by Bebbington and Cronin (2011) has discounted earlier studies that suggested spatio-temporal clustering in the AVF. Instead the spatial and temporal aspects appear independent; hence the location of the last eruption provides no information about the next location.

Previous AVF eruptions have typically been small in volume (<0.1 km^3, Allen and Smith 1994), However the last two eruptions, Rangitoto (2 km^3) and Mt. Wellington (0.17 km^3) are two of the largest in volume, suggesting a possible change in future eruptive behaviour (Lindsay 2010).

The eruption style during vent opening is typically phreatomagmatic, due to rising magma interacting with groundwater and/or seawater (if a vent occurs in the ocean). Some eruptions cease after this stage, leaving broad maars or explosion craters typically 1-2 kilometres in diameter. Where the eruptions continue beyond this stage, subsequent activity is of magmatic Hawaiian style, which produces scoria cones and lava flows. The explosive phreatomagmatic AVF eruptions have generated volcanic hazards such as base surges, a type of pyroclastic density current (denser-than-air flows which can travel at 200-300 km h^{-1} and be >200 °C; Browne 1958, Belousov et al. 2007), shock-waves and ballistics (material >64 mm erupted from the vent) which are highly destructive to areas up to 3 kilometre radius of the vent. Secondary hazards, such as earthquakes, tephra fall and gas release, would also be noticed throughout the entire region. The footprint of these hazards from previous AVF eruptions and those of other analogous volcanoes have been used to calculate evacuation zones for future events. Current contingency and mass evacuation plans call for areas of 3 kilometres ("Primary Evacuation Zone") and 5 kilometres ("Secondary Evacuation Zone") radius from the erupting vents to account for base surges (Beca Carter Hollings and Ferner Ltd. 2002, Auckland CDEM Group 2008b, Auckland CDEM 2013).

Finally, It is likely that civil authorities and area residents will only be provided with at most a few weeks and as little as a few days of warning time prior to an eruption (Beca Carter Hollings and Ferner Ltd.

2002). This is based on the expected fast magma ascent rates (1-10 cm s^{-1}) that basaltic volcanic fields are known to exhibit (Blake et al. 2006, Sherburn et al. 2007). Early detection of precursory activity is therefore critical. However, the factors contributing to a decision by emergency management officials to call a mass evacuation bring significant levels of uncertainty to mass evacuation planning in Auckland. Furthermore, the exact vent area is likely to be unknown until shortly before outbreak (Blake et al. 2006). One attempt to address this uncertainty was presented by Sandri et al. (2012), who developed a cost-benefit analysis model for evacuation planning by weighing the cost of issuing evacuation warnings for geographic areas (represented by lost work potential, warning costs, movement costs etc.) against the benefit of evacuating (represented by the number of lives saved). When the benefits of evacuating a certain area exceed the associated costs, an evacuation is deemed warranted. The point of changeover is the ideal boundary to use as the time to call for evacuation. Sandri et al. (2012) calculated probability threshold values for a range of magnitudes: small effusive, moderate phreatomagmatic and large phreatomagmatic eruptions. Evacuation radii for these three scenarios were established at 3.5, 5 and 8 kilometres respectively. These are somewhat consistent with the primary (3 km radius) and secondary (5 km radius) evacuation zones in the AVF contingency plan, but crucially suggest an additional larger radius of 8 km should be considered in evacuation demand analysis. The evacuation area was also found to change in size with time in the lead-in period, due to a reduction in the uncertainty in the vent location and increase in the probability of an eruption. Thus, there is a trade-off between these two factors (area and time) that dictates which cells must be evacuated, and when (Sandri et al. 2012). Given the uncertainty in vent location (and the subsequent need to wait until close to outbreak to define the evacuation area) it is likely that the evacuation will need to be carried out quickly, i.e. within the 48-hour evacuation time required by civil authorities.

DATA SOURCES

Two primary types of data are used in this study: population data and infrastructure data. Population data from Statistics New Zealand, which provides data in its *Census of Population and Dwellings,* was sourced from the results of the 2006 census. Census areas are arranged hierarchically with *regions*representing the most extensive geographical areas, and *meshblocks* representing the smallest areas. The more detailed meshblock scale is used for determining evacuation demand in this research. We also use *neighbourhood areas*; defined here as areas bounded by major and arterial roads, motorways, or saltwater inlets. Census data utilised in the research includes night-time population, average household sizes, ages, income levels, vehicle availability per household, and business demography. Data regarding school attendance was obtained from the Ministry of Education. The majority of infrastructure data employed was provided by the former Auckland Regional Council and were current as of October 2009. This included geospatial road data files, which contained comprehensive attributes for the Auckland region such as road hierarchy, designated speed limits and directionality. In order to facilitate macro-evacuation studies covering areas outside Auckland, major and arterial road and motorway data for Northland and Waikato, produced by NZ Open GPS Maps (2009), was added to the database. Since our study a new Census has been carried out in New Zealand, and we note that it would be worthwhile repeating this study with post-2009 infrastructure data together with the new (2013) Census data once it becomes available.

METHODS

Constraining the Study Area

Various size estimates have been employed to represent the size of the AVF. For this study we adopt the same bounding limits used by Lindsay et al. (2010), reflecting the continuity of the underlying geology and recognising the possibility of eruptions taking place outside the current extent of the AVF. However, we employ an ellipse rather than a rectangular area since established volcanic fields are more often ovular in shape (Spörli and Eastwood 1997). The resulting ellipsoid (Figure 2) represents the large conservative 'future geologic extent' of the AVF and is used as the theoretical boundary of the field in this study. Subsequent to our study, further work by Le Corvec et al. (2013a,b) has provided new insights into a smaller quantitative boundary for the AVF. Were such a smaller, less-conservative ellipsoid representing the current extent of the AVF used, all of our calculated probabilities would be much larger, as the total area (much of which is water) is reduced. Also, the chance of zero evacuation demand would likely be minimal, particularly if evacuations resulting from tsunami generated by offshore eruptions were incorporated into the study.

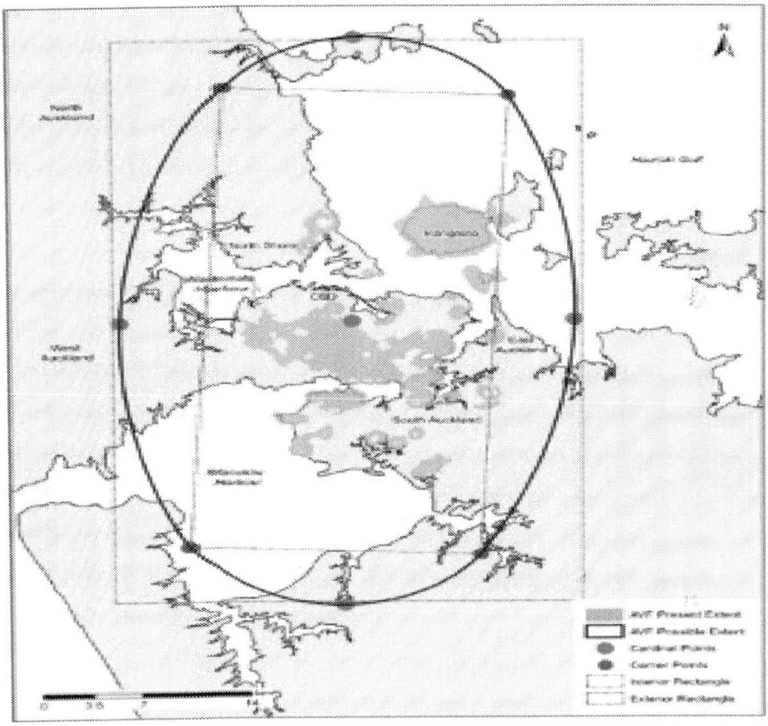

Figure 2: Geometry of the Auckland Volcanic Field. The ellipsoid represents the possible future geologic extent of the AVF and is used in this study as the presumed theoretical boundary of the field.

Population Evacuation Demand

Traditional census data in New Zealand reports the population according to their night-time residence but does not cover daytime population figures. Business demographics data from Statistics New

Zealand (2008) and school enrolment data for 2009 provided by the Ministry of Education were thus used to estimate Auckland's daytime population. Using employing units with a Goods and Services Tax turnover of $30,000 or more, there were 593,276 people employed *in* Auckland and 601,638 people employed *from* Auckland, suggesting that roughly 1.4% of Aucklanders leave the region for work. The number of students enrolled in Auckland schools was 244,449. As the Ministry of Education do not code their data to show where students reside, a proxy value based on age cohorts from the census data was obtained (sum of age 5-9, 10-14 and 15-19 cohorts). However, as not all people in this range attend school, the proxy value (281,460) overestimates the enrolment generation by 15%. Equation 1, which was used to generate daytime population, used the night-time population as a baseline measure and added the inflows of workers and students, then subtracted their outflows:

Day pop = Census night pop.+ workers in + school in – workers out – school out (1)

Using the night-time and daytime statistics together allows the magnitude of diurnal movement of Auckland's populations to be calculated. However, although diurnal trends in population are thoroughly assessed in this study, these are largely based on scenarios for a typical weekday. Further work to obtain detailed estimates of population distribution over space and time would greatly assist evacuation planning in Auckland.

To calculate population evacuation demand, we assumed that all evacuees in the danger areas would evacuate, and that there is no shadow evacuation outside of those areas. While these assumptions are simplistic and unlikely to represent the true human dynamics of a mass evacuation, they allow for the creation of baseline values. We also assumed an equal probability distribution of a new vent forming anywhere in the AVF and calculated population evacuation demand (at 100-metre resolution) during day and night for every point in the AVF. This was accomplished first by determining the population density of each census mesh block and transforming the densities into aerial-based population values in ArcGIS (2009). Then, the AVF extent ellipsoid was buffered by the evacuation radius values. This was done to reflect the reality that eruptions occurring on the edge of the AVF will impact those within the evacuation radius even if they are outside the AVF boundary. In the final step of the process, the population evacuation demand was calculated for each 100 m^2 cell by summing the values of the individual cells within designated neighborhood distances. The three evacuation radii proposed by Sandri et al. (2012), 3.5, 5 and 8 kilometres, were used as the neighbourhood distances in successive calculations, thus producing three models. Each model was clipped to match the extent of the AVF study area and contour lines were created at intervals of 50,000 people to enable improved visualisation of the data. The cells were also reclassified into 50,000 unit intervals, allowing the results to be displayed numerically as 11 classes. For ease of comparison, we focus our results on the 5 kilometre evacuation radii, which is also the larger of the 'most probable' evacuation radii used in the current AVF contingency plan.

In addition to total number of evacuees, the low-mobility population evacuation demand is also of concern for public authorities, particularly as they may rely on public transportation. Both census and school enrolment data were used to determine the spatial and numeric extent of the population without a vehicle for each meshblock using Equations 2 and 3 for night and day respectively.

Low mobility pop (night) = no motor vehicle households × average household size (2)

Low mobility pop (day) = (#motor vehicle households × av. household size) + school enrolment (3)

It was assumed that no school children drive to school. Therefore, all school children are considered low-mobility during the day, but not at night when they return home to their families. Population evacuation demand figures for the low-mobility population were calculated using a similar technique as for the general population, the major alteration being that low-mobility meshblock statistics, rather than total meshblock statistics, were adopted as the inputs. The statistical results were reclassified into 12 classes at 10,000 unit intervals.

Network Analysis

Two types of network analysis were conducted in this study: Spatial analysis to determine the geographic functionality of different transport hubs and evacuation destinations; and vulnerability analysis for the four major motorway bridges which are integral to Auckland's transportation network. ArcGIS (2009) was employed, with constraining values such as movement restrictions and distances adopted to calculate populations that reside within easy reach of different transportation assets. Such assets included public transport stops (bus stations, ferry terminals and train stations), boat ramps, marinas and welfare facilities. It should be noted that public transportation in Auckland is more limited at certain times such as at night and at weekends, and that the functionality of such transport depends on the reliability of its workers and their ability and willingness to travel to work, no matter what the timing.

All roads and footpaths in Auckland were used to create the network dataset and the daytime and night-time populations for meshblocks within transport service areas were summed. We chose 30 minutes as the maximum time that evacuees would be willing to walk to a departure point and, based on a consistent walking rate of 5 km h^{-1}, a time interval representing accessibility was set correspondingly. As most boat owners access boat ramps using motor vehicles, we adopted an accessibility time interval of 15 minutes, based on average rush hour speeds for different road types (Auckland Regional Council 2009b) for this asset type.

The four primary motorway bridges that form essential links in Auckland's transportation network are the Auckland Harbour Bridge and Northwest Motorway Bridge at the north end of the isthmus, and the Southern Motorway Bridge and Mangere Bridge at the south end (Figure 1). For the purpose of this study, the vulnerability of these bridges was based on the likelihood of the structures lying within each of the designated evacuation radii. Geospatial buffers were created for each bridge to match the extent of the three evacuation radii and these were subsequently clipped to the extent of the AVF. The resulting areas were then divided by the total area of the AVF. This provided the probability that any bridge would intersect the evacuation zone and thus would likely be damaged or made unusable by the eruption. The directional vulnerability (i.e. that either bridge in one direction would be impacted) and combined vulnerability of any of the four bridges being impacted was also assessed. We achieved this by merging and intersecting the existing bridge-evacuation radii areas with one another in ArcGIS (2009) and dividing the new areas by the total area of the AVF.

Evacuation Vulnerability

Evacuation vulnerability was analysed at both the micro- and macro-scale. In order to assess evacuation difficulty at the micro-scale, we employed a method using population to exit capacity (P/EC) ratios, first pioneered by Cova and Church (1997). Neighbourhoods, which are commonly used for transportation modelling, were used as a key unit of analysis for this study. This was deemed appropriate, as evacuation zones need to be easily differentiated by the public and neighbourhood boundaries (i.e. major and arterial roads, motorways and saltwater inlets) are easily distinguishable by eye. Although there is no mention of boundary delineation in the current mass evacuation plan for Auckland, clear geographic and functional features were used to determine evacuation zones during Exercise Ruaumoko (see Lindsay et al. 2010). Therefore we used saltwater inlets, motorways, and major and arterial roads to define the neighbourhoods and thus evacuation zone boundaries in this study. This ensures that the majority of traffic will flow towards the periphery of the neighbourhood from the onset of evacuation, rather than to a point within the zone, and then outwards. After determining the neighbourhood zones, the points where each road intersected the neighbourhood boundary were designated as exits. Exits were coded according to the neighbourhood they serviced and their capacity, and half of all one-way streets were removed, as the direction of flow for these was indiscernible. The total exit capacity for each neighbourhood was generated and daytime and night-time populations were calculated based on the neighbourhood areas. Dividing the populations by the total exit capacities thus produced the P/EC ratios.

To statistically test correlations between P/EC ratios and neighbourhoods with different constricting boundary types (i.e. motorways and water), we employed a two-tailed t-test with a 95% confidence interval. The null hypothesis stated that there was statistically no difference between the mean of the first data set (μ_1) and the second data set (μ_2). The alternative hypothesis claimed that the difference is statistically significant. Four t-tests were performed. The first three assessed each of the two constricting boundary types individually and in combination against the neighbourhoods with neither boundary type. The final assessment compared neighbourhoods with neither boundary type against all data values in aggregate.

As most evacuees in developed countries prefer to evacuate by private vehicle, we assumed that each household leaves in one vehicle, and, following the same procedure used to calculate P/EC ratios, household to exit capacity (HH/EC) ratios were calculated. Sensitivity analysis on this assumption was also conducted by assuming that all available vehicles per household are used during evacuation. This was achieved first by using Equation 4, followed by the same method that was used to calculate P/EC ratios, thus determining Car/EC ratios.

$$\text{Cars} = {}^{n}\text{HH (one car)} \times 1 + {}^{n}\text{HH (two cars)} \times 2 + {}^{n}\text{HH (threeormorecars)} \times 3 \qquad (4)$$

where n_{HH} represents the number of households in each census meshblock.

Equation 4 slightly underestimates the number of cars per meshblock, because it is based on the assumption that no household has more than three cars, which some likely do, though data that would resolve this is unavailable at present.

TransCAD models were built to assess the feasibility of overland movements along the regional roadway network in Auckland and to test various independent variables on a macro-scale to determine the degree to which they impact the network clearance time (TransCAD, developed by Caliper Corporation, is a GIS with highly specialised transportation modelling capabilities). Such models dissect the transportation network as a set of links (roads) and nodes (intersections or endpoints). Centroids are a special set of nodes representing the geographic centre of a Traffic Analysis Zone (TAZ). Centroids begin and end every modelled trip and therefore represent both origin (neighbourhood) and destination (welfare facility) locations. They are connected by the Auckland road network, which is generalised to improve processing. A generalised road network also has the advantage of more realistically modelling evacuation flows because the public's knowledge of the road network will be imperfect and they are most likely to remain on more familiar routes such as major roads and motorways (Tomsen 2010).

Designated Auckland Civil Defence and Emergency Management welfare facilities were chosen as the destination locations for intra-regional evacuation studies on the macro-scale. This is because, in an ideal situation, all evacuees would go to such a facility to register with emergency management officials before heading to their final destination. For inter-regional analysis, two major welfare facilities outside the Auckland region, one in Hamilton (south) and the other in Whangarei (north) were chosen as the destination locations (Figure 1). For the purpose of this study, the road network outside of the Auckland region was generalised in two ways: motorways only, and motorways in combination with major arterial roads. However, within the Auckland region, all major and arterial roads were utilised, forming the intricate lattice of TAZ boundaries. In macro-evacuation models, TAZs do not contain intra-zonal roads; therefore, secondary roads in Auckland were removed by definition. Vehicle capacity values were assigned to each road type in the models. As capacity values were not available for Auckland, estimates provided in the Brisbane Strategic Transport Model (Brisbane City Council 2007) were adopted. These values range from 1,100 vehicles per lane per hour for arterial roads to 2,100 vehicles per lane per hour for 4-lane motorways. The number of centroid connectors determines the number of neighbourhood exits that are utilised. While this has no impact on intra-zonal movement, the effects of congestion begin as soon as the first non-centroid node is reached. To determine the effect of this congestion, the number of centroid connectors was varied as a form of sensitivity analysis. Other forms of sensitivity analysis included modification for time of day, evacuation radii, vent locations, welfare facility locations, shadow evacuations and various levels of road network detail.

TransCAD models were run for AVF vent locations with minimum and maximum population evacuation demand values using the 5 kilometre evacuation radius (determined from results of the micro-vulnerability work). Since there were multiple locations with no evacuation demand, vent locations furthest from the coast in both Waitemata and Manukau Harbours were selected as the minima. The maximum population evacuation demand value was generated from a vent location one kilometre south of Mount Eden. Additionally, the vent used for Exercise Ruaumoko, 0.7 kilometres west of Mangere Bridge in the Manukau Harbour (see Lindsay et al. 2010), was also selected as a scenario.

Outputs from TransCAD models include total flows over the entire network, Volume-Over-Capacity (VOC) ratios and total network clearance times. Flows indicate the total volume of traffic passing over each network link, which in turn provides data on which links are most heavily used and thus the most critical during evacuations. Total flows over the entire network equalled the total number of vehicles involved in the evacuation. Flows alone, though useful in visualising movement, cannot predict congestion, because they do not account for the capacity of the road segments. VOC ratios excel at this and are a common measure of assessing movement difficulty (Cova and Church 1997, Church and Cova 2000, Shulman 2008, Marrero et al. 2010). VOC ratios examine the volume of traffic on each lane of the road divided by the lane's capacity (similar to Car/EC ratios, except VOC ratios are based on links rather than zones). High ratios are indicative of congestion and when the VOC approaches 1, the velocity of traffic approaches 0. The clearance time (i.e. the time required to evacuate Auckland) was calculated for different inter- and intra-regional evacuation scenarios by linking the network travel times to the original road network.

TransCAD models, like all other evacuation models, rely on the accuracy of the data they employ and the underlying assumptions within the models. While the baseline datasets of population and road infrastructure are accurate, the derivative datasets used to produce origin-to-destination functions are highly dependent on the validity of numerous assumptions. Some assumptions, including evacuation compliance, shadow evacuation and vehicle use rates can be measured indirectly through questionnaires and tested in models with sensitivity analysis; others, such as the destination of evacuees, can only be presumed and aggregated. Further assumptions are built into the model and cannot be altered. For instance, TransCAD models rely on a constant road network and consistent driver interaction. This ignores the possibility of traffic accidents, network alterations due to ongoing damage, and the possibility of panic or "road rage" on the part of drivers. A further limitation with TransCAD is its click-and-run start-up method, which prevents evacuees from being added to the model at successive time intervals. Essentially, the number of evacuees the model begins with is how many it processes to completion. No variation in preparation time is accounted for. Models are also limited to egress population movements, disregarding background traffic. Background traffic is a legitimate concern if little forewarning is provided (Urbanik 2000); yet, when sufficient forewarning is provided (as is expected with an AVF eruption), background traffic is unlikely to pose significant issues (Lindell and Prater 2007).

RESULTS

Population Evacuation Demand

The total night-time population of Auckland, based on the raw census data (2006) was 1,264,011. The daytime population based on typical worker and student migration was calculated to be 1,218,638. Thus, a loss of 3.6% of the population is noted in the statistics and used as the basis of further analysis for this study. Geographic analysis of diurnal movement in Auckland showed that areas of greatest population gain during the day were business areas, whereas areas of greatest population loss were residential areas. The Central Business District (CBD) had nearly three times the population influx of any of Auckland's other 238 neighbourhoods and thus is likely to be much more difficult to evacuate during the day than night.

The numeric results of the population evacuation demand model outputs are presented in Table 1. Two outputs of this model, utilising the 5 kilometre evacuation radii and the two time variables, are shown in Figure 3. The percentage values reflect the proportion of the cells in the AVF that fall into each category. Because this study employs an equal probability distribution of an eruption occurring at any location in the AVF, the percentage values also represent the likelihood that a future AVF eruption will generate the number of evacuees listed in each class. For instance, using a 5 kilometre evacuation radius, there is a 56.4% chance that 1-50,000 evacuees will result from a daytime eruption.

Table 1: Population evacuation demand statistics

Evacuation Demand	Day			Night		
	3.5 km	5 km	8 km	3.5 km	5 km	8 km
0	11.8%	3.9%	0.0%	11.7%	3.9%	0.0%
1 – 50,000	65.5%	56.4%	37.3%	62.3%	50.8%	32.4%
50,000–100,000	18.7%	18.0%	13.0%	23.6%	20.9%	13.7%
100,000-150,000	2.1%	11.2%	12.1%	2.3%	19.5%	11.7%

150,000-200,000	1.2%	5.7%	10.4%	0.0%	3.8%	13.4%
200,000-250,000	0.8%	2.5%	5.9%	0.0%	1.1%	9.7%
250,000-300,000	0.0%	1.7%	5.0%	0.0%	0.0%	7.2%
300,000-350,000	0.0%	0.6%	4.9%	0.0%	0.0%	5.8%
350,000-400,000	0.0%	0.0%	4.3%	0.0%	0.0%	6.0%
400,000-450,000	0.0%	0.0%	4.7%	0.0%	0.0%	0.0%
450,000-500,000	0.0%	0.0%	2.5%	0.0%	0.0%	0.0%
Average	27,210	55,077	137,317	27,088	55,346	134,268
Minimum	0	0	6	0	0	6
Maximum	237,405	327,937	505,677	131,841	239,895	397,549

Tomsen *et al.*

Tomsen *et al. Journal of Applied Volcanology* 2014 **3**:6, doi:10.1186/2191-5040-3-6

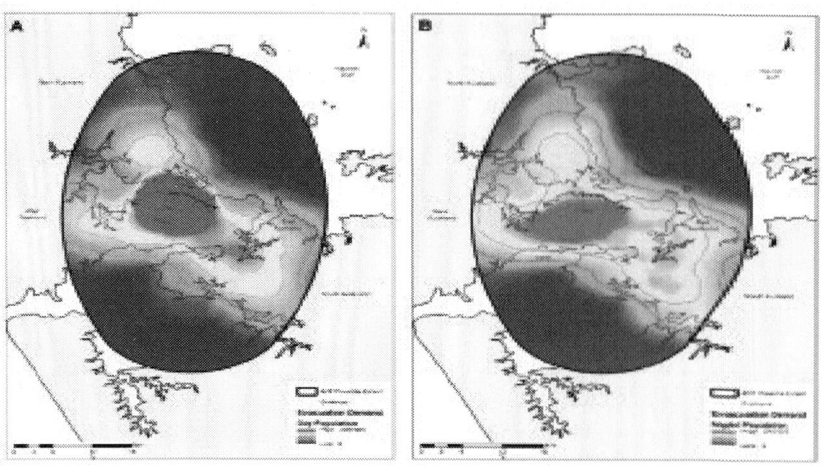

Figure 3: Daytime (A) and night-time (B) population evacuation demand for a 5 kilometre evacuation radius. Note the different scales used for the daytime and night-time evacuation demand colour-ramps.

Larger evacuation radii were found to substantially increase the average and maximum population evacuation demands and diminish the initial spike in evacuation demands present for smaller radii. During daytime, evacuation demand values are high around the CBD area but become more widely dispersed as the radius increases or during night-time. Correspondingly, the maximum evacuation demand values are substantially smaller for night-time than daytime evacuations (the average maximum daytime value is 357,000 using all three radii, while the average maximum night-time value is 256,000). The results show that there are some locations in the AVF that would generate no evacuation demand, because the vents are located sufficiently far offshore (reflected as zero values in Table 1). Using the 3.5 kilometre radius, there is a 12% chance that an eruption will generate no evacuees. Using the 5 kilometre radius, this figure is reduced to 4%, and with an 8 kilometre radius to 0%, though the minimum values remain quite low. However, it should be noted that we do not consider evacuation need due to tsunami generated by offshore volcanic eruptions in our study. Such an event may result in increased evacuation demand in coastal areas.

Low-mobility Population Evacuation Demand

Based on school enrolment figures and households with no vehicles, there are 321,162 people with low-mobility in Auckland during the day and 76,713 people at night. This represents 25.4% and 6.1% of Auckland's total population, respectively. These calculations may overestimate the daytime low-mobility population, since low-mobility groups increasingly use social networks to obtain rides with friends or neighbours rather than relying on public transportation (Lindell and Prater 2007). Additionally, an unknown percentage of school children come from low-mobility households and thus would be double-counted during the day, and some parents may pick up children from school during an evacuation. A proportion of older students may also drive to school, although we expect that this is more prevalent in rural communities than urban centres such as Auckland. All of these variables however, are somewhat counterbalanced by an underestimation of the number of households with no access to motor vehicles, which result from the meshblock level census data for car ownership providing no data for some areas (due to small sample sizes being deemed confidential by Statistics New Zealand). Also, other low-mobility groups such as the hospitalised and institutionalised are omitted from the calculations. The former Auckland Regional Council calculated that 7.4% of Aucklanders have no access to a motor vehicle - a figure comparative to the 6.1% value used in this study, justified as the council's calculations are based on a regional rather than meshblock scale (Auckland Regional Council 2009a). The statistical results for the low-mobility population evacuation demand are presented in Table 2 and two outputs of the geographic model utilising the 5 kilometre evacuation radii and two time variables, are shown in Figure 4, overlaid atop the ferry and rail station stops.

Table 2: Low-mobility population evacuation demand statistics

Evacuation Demand	Day			Night		
	3.5 km	5 km	**8 km**	**3.5 km**	**5 km**	**8 km**
0	20.6%	11.8%	4.0%	21.0%	11.8%	4.0%
1-10,000	47.7%	40.2%	26.9%	78.7%	77.7%	59.3%
10,000-20,000	22.5%	15.0%	10.9%	0.4%	10.5%	22.9%
20,000-30,000	7.3%	14.1%	8.6%	0.0%	0.0%	13.5%
30,000-40,000	1.9%	12.7%	9.2%	0.0%	0.0%	0.4%
40,000-50,000	0.0%	4.5%	10.0%	0.0%	0.0%	0.0%
50,000-60,000	0.0%	1.3%	7.1%	0.0%	0.0%	0.0%
60,000-70,000	0.0%	0.5%	7.0%	0.0%	0.0%	0.0%
70,000-80,000	0.0%	0.0%	5.2%	0.0%	0.0%	0.0%
80,000-90,000	0.0%	0.0%	4.0%	0.0%	0.0%	0.0%
90,000-100,000	0.0%	0.0%	5.7%	0.0%	0.0%	0.0%
100,000-110,000	0.0%	0.0%	1.5%	0.0%	0.0%	0.0%
Average	6,945	13,993	34,672	1,722	3,473	8,652
Minimum	0	0	0	0	0	0
Maximum	39,283	66,629	108,455	11,375	19,073	31,610

Tomsen *et al.*

Tomsen *et al. Journal of Applied Volcanology* 2014 **3**:6, doi:10.1186/2191-5040-3-6

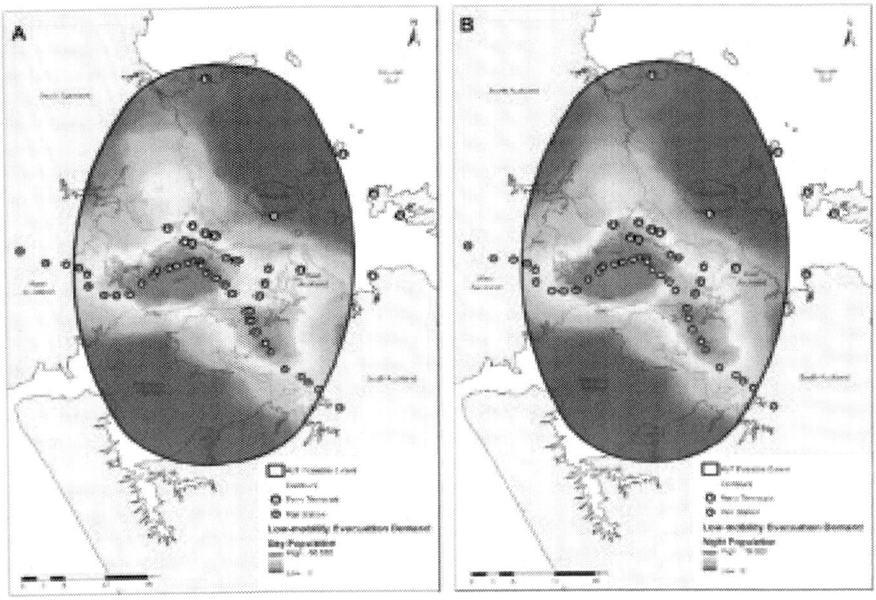

Figure 4: Daytime (A) and night-time (B) low-mobility population evacuation demand for a 5 kilometre evacuation radius. Note the different scales used for the daytime and night-time evacuation demand colour-ramps.

Network Analysis

Geospatial network analysis showed that the bus network, with a high number of stops and wide geographic distribution, was accessible to the majority of the Auckland population. Based on a standardised walking rate of 5 km h^{-1}, over 92% of Aucklanders can reach a bus stop within a 30 minute walk regardless of the time of day. Other networks such as trains and ferries, were significantly less accessible; The number of Aucklanders who can reach train stations or ferry terminals within a 30 minute walk were 37% (night-time) to 47% (daytime) and those who can reach ferry terminals in the same time were 5% (night-time) to 12% (daytime). The percentage of people living within a 15-minute drive, based on typical rush hour traffic speeds (Auckland Regional Council2009b), of the nearest marina is relatively low (31-37%), whereas the percentage within 15 minutes of a boat ramp is high (90-92%).

Individual vulnerable bridge probability values for the three evacuation radii range from 3.1% to 14.7%. Combined bridge vulnerability statistics are listed in Table 3. The maximum probability of either one of the north or south bridges being impacted nearly doubles as the radius is increased. For the 5 kilometre evacuation radius, there is a 1.7% chance that both north bridges, and a 1.5% chance that both south bridges, will be impacted but 0% chance that three bridges will be impacted by the eruption (Figure 5). In the event of an 8 kilometre evacuation radius however, there is a 1.7% chance that three bridges could be simultaneously impacted. It should be stressed that if the overall size of the AVF were reduced to include a more restricted (less conservative) ellipse around the existing volcanic cones (Figure 1), the probability of each event described above would be dramatically increased.

Table 3: Combined bridge vulnerability statistics

Bridges	**3.5 km**	**5 km**	**8 km**
Any	12.7%	21.5%	38.7%
Either North	7.3%	11.9%	21.8%
Either South	5.4%	9.6%	20.6%
Both North	0.2%	1.7%	7.7%
Both South	0.2%	1.5%	7.3%
NW, Harbour & Mangere	0.0%	0.0%	1.7%

Tomsen *et al.*

Tomsen *et al.* *Journal of Applied Volcanology* 2014 **3**:6, doi:10.1186/2191-5040-3-6

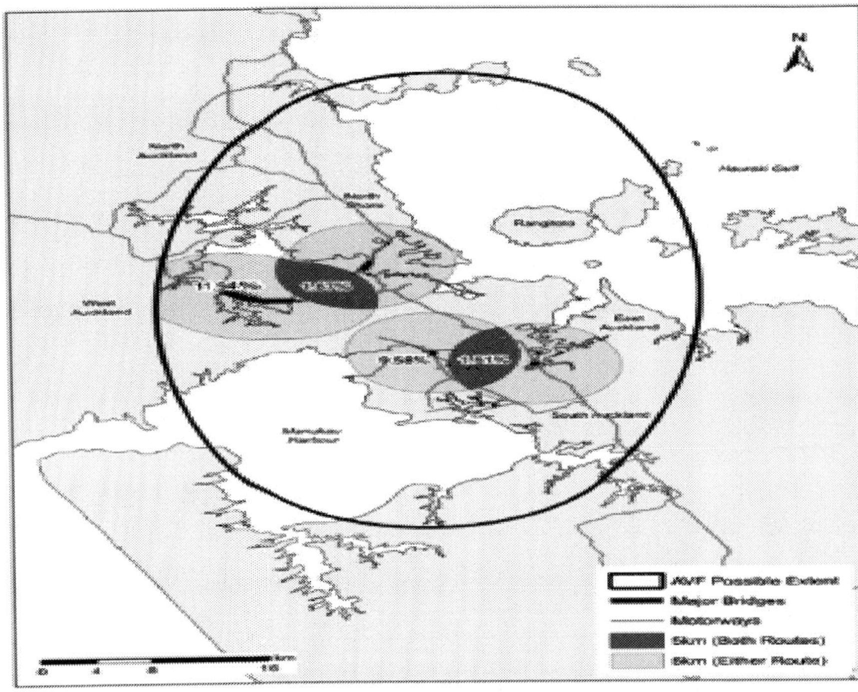

Figure 5: Directional bridge vulnerability for a 5 kilometre evacuation radius. This shows the probabilities of either bridges and both bridges to the north or south being impacted by an eruption. Note that if a smaller, less-conservative ellipse representing the extent of the AVF were used in this study, the probability of each event would be higher.

Micro-scale Evacuation Vulnerability

Diurnal changes in population evacuation demand are reflected in the calculated P/EC ratio values (population/exit capacity, described above). Although no data exists for threshold P/EC ratios for different roads in Auckland, a comparison of values provides a first-hand relative assessment of potential congestion areas. There is a high concentration of areas with elevated P/EC ratios during the day in and around the CBD, which at night is dramatically reduced. The reverse is true for many outlying suburbs, particularly those in West Auckland and the North Shore, which experience much higher ratios at night. This makes intuitive sense as many people travel from the suburbs to the CBD for work each day. Table 4 lists the five neighbourhoods with the highest and five neighbourhoods with the lowest P/EC ratios. Two of the five neighbourhoods with the greatest net population gain during the day, the CBD and Auckland Airport, also have the two highest P/EC ratios during the day (1,819 and 1,574 people per exit lane, respectively). Night-time P/EC ratios for many of these neighbourhoods are significantly lower. Neighbourhoods with the greatest net population losses during the day also have some of the highest night-time P/EC ratios. Overall, the highest P/EC ratios during the night are far lower than the highest P/EC ratios during the day. Only two neighbourhoods, both in south Auckland, have P/EC ratios over 1,000 at night.

Table 4: Neighbourhoods with highest and lowest Population/Evacuation Capacity (P/EC) ratios for day and night

DAY*				NIGHT*			
Rank	**Neighbourhood**	**P / E C Day**	**P/EC Night**	**Rank**	**Neighbourhood**	**P / E C Night**	**P/EC Day**
1	Puhoi	10.3	28.5	1	Druces West	4.3	710.9
2	Gardens East	11.3	42.9	2	Manukau South	6.5	361.3
3	Glenbrook	15.1	34.6	3	Sylvia Park	10.3	1292.4
4	Ormiston North	18.9	41.9	4	Hampton Park	10.8	595.1
5	Greenhithe	23.6	84.2	5	Pukekohe North	16.2	123.9
234	Sylvia Park	1292.4	10.3	234	Wattle Farm	778.6	450.0
235	Rosedale	1362.5	267.0	235	Sunnynook	842.1	376.3
236	Hingaia	1567.9	685.3	236	Gulf Harbour	886.0	448.3
237	Akld Airport	1574.3	89.9	237	E Howick Beaches	1000.6	653.8
238	Akld CBD	1818.9	447.0	238	Mt. Mangere	1291.4	753.4

*The highest and lowest ranked P/EC ratios are shown in bold font. These are compared with the corresponding P/EC ratios for the contrasting diurnal time shown in normal font. Tomsen *et al*.

Tomsen *et al. Journal of Applied Volcanology* 2014 **3**:6, doi:10.1186/2191-5040-3-6

When graphically comparing daytime and night-time P/EC ratios, the prevalence of neighbourhoods with lower P/EC ratios is very clear. Based on the Brisbane Strategic Transport Model (Brisbane City Council 2007) access roads, such as those connecting neighbourhoods to primary roads, can facilitate the movement of up to 600 vehicles per hour. Therefore P/EC ratios > 600 are very likely to cause congestion and the value was used as a benchmark. In our study, the number of neighbourhoods with P/EC values greater than 600 is very small both at day and night (Figure 6).

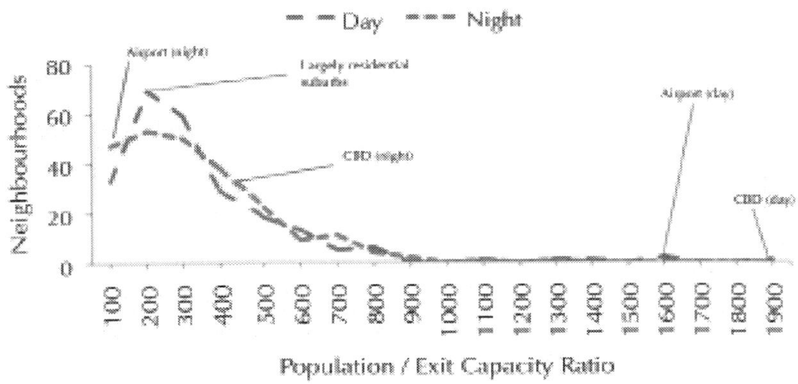

Figure 6: Daytime and night-time Population/Exit Capacity (P/EC) ratios. The P/EC ratios for key neighbourhoods described in the text are shown on the graph.

Geographic features are of great consequence when assessing P/EC ratios. Neighbourhoods with both motorway and water boundaries have higher P/EC ratios. This is to be expected, as boundaries work as constraining geographic features limiting the number of exits. The results of statistical analysis of neighbourhood boundaries are shown in Table 5. All p-values, with two exceptions, were less than 0.05 and thus statistically significant.

Table 5: Statistical t-test comparison of neighbourhood boundary types

μ_1	μ_2	**p-value (day)**	**p-value (night)**
Both	Neither	0.020	0.012
Motorway Only	Neither	0.020	0.632
Water Only	Neither	0.623	0.004

Neither	All	0.020	0.034

Tomsen *et al.*

Tomsen *et al. Journal of Applied Volcanology* 2014 **3**:6, doi:10.1186/2191-5040-3-6

These results show that there is a significant difference between neighbourhoods with only motorways as boundaries and neighbourhoods with neither water nor motorways as boundaries during the day (i.e. when people are at work). However, at night there is no statistically significant difference. Contrastingly, neighbourhoods with only water boundaries are significantly different during the night (i.e. when those people return home), but not during the day. This is likely due to more residential neighbourhoods being located close to water features which provide both aesthetic and recreational value. Neighbourhoods containing businesses are more likely to occur close to motorway on- and off-ramps, which enable easy access for workers and goods. Therefore, when either of these neighbourhood types reach their peak population, they usually become more difficult to evacuate.

The numeric results of the HH/EC ratio and Car/EC ratio calculations are presented in Table 6. Results of both studies were found to be similar to those of the night-time P/EC ratios. HH/EC ratios were generally one third that of the P/EC ratios; this correlates with the fact that the mean Auckland household has three residents (Statistics New Zealand 2006). Car/EC ratios were around half as large as P/EC ratios, which correlates with there being roughly twice as many people as cars in Auckland (Statistics New Zealand 2006, New Zealand Transport Agency 2009). The minima Car/EC ratios at or bordering zero are suspect and likely due to the fact that such areas have very low residential populations. When compared to P/EC ratios and Car/EC ratios, HH/EC ratios represent the best-case scenario, as they minimise the overall number of evacuating units and thus the ratio, thereby producing fewer units per exit lane.

Table 6: Neighbourhoods with highest and lowest Household/EC Ratios and Car/EC Ratios

Household/Evacuation Capacity (HH/EC)*					Car/Evacuation Capacity (Car/EC)*				
Rank	**Neighbourhood**	**HH/EC**	**Car/EC**	**P / E C Night**	**Rank**	**Neighbourhood**	**C a r / EC**	**H H / EC**	**P / E C Night**
1	Manukau South	0.5	0.7	6.5	1	Druces West	0.0	2.1	4.3
2	Druces West	2.1	0.0	4.3	2	Sylvia Park	0.0	3.2	10.3
3	Sylvia Park	3.2	0.0	10.3	3	Manukau South	0.7	0.5	6.5
4	Hampton Park	4.9	6.2	10.8	4	P u k e k o h e North	5.5	5.3	16.2
5	Hobsonville	5.2	9.6	16.7	5	Hampton Park	6.2	4.9	10.8
234	West Tamaki	246.4	361.4	771.5	234	Hingaia	454.5	229.0	685.3
235	Sunnynook	280.1	472.9	842.1	235	Sunnynook	472.9	280.1	842.1
236	Gulf Harbour	313.1	540.0	886.0	236	Gulf Harbour	540.0	313.1	886.0
237	E Howick Beaches	328.2	628.1	1000.6	237	E Howick Beaches	628.1	328.2	1000.6
238	Mt. Mangere	411.6	643.2	1291.4	238	Mt. Mangere	643.2	411.6	1291.4

*The highest and lowest ranked Household/EC ratios and Car/EC ratios are shown in bold font. These are compared with the corresponding Car/EC ratios and Household/EC ratios, and also P/EC ratios, which are shown in normal font.

Tomsen *et al.*

Tomsen *et al.* *Journal of Applied Volcanology* 2014 **3**:6, doi:10.1186/2191-5040-3-6

Macro-scale Evacuation Vulnerability

For inter- and intra-regional evacuations, flows often exceeded 50,000 vehicles along motorway sections, but rarely exceeded 20,000 vehicles for arterial and major roads. Symbolising flows along different routes by means of varying line thickness allows these differences in flows to be visualised (Figure 7).

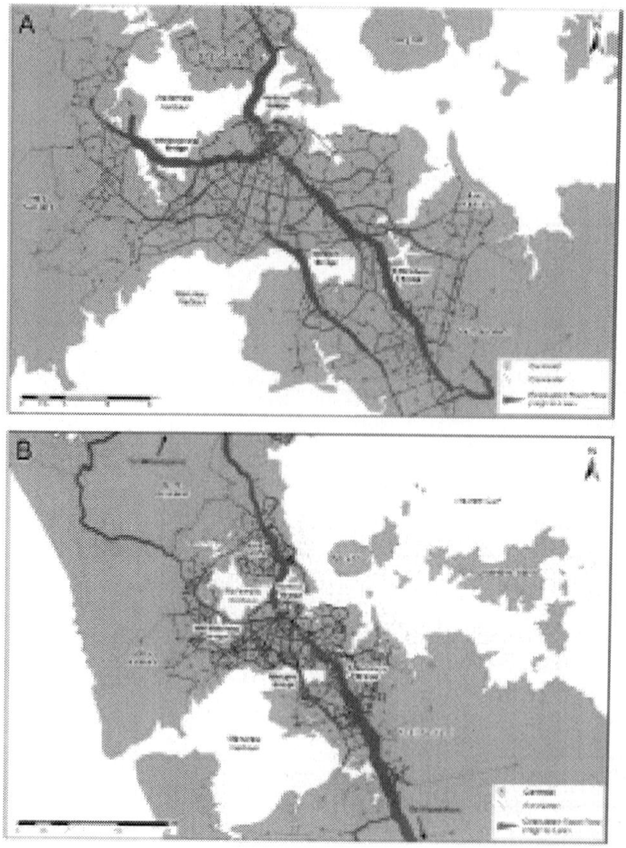

Figure 7: Symbolised graduated road flows for intra-regional (A) and inter-regional (B) evacuation. The total flows over the entire network equals the total number of vehicles involved in the evacuation. Along motorway sections, flows often exceed 50,000 vehicles. Flows for other arterial and major roads rarely exceeded 20,000 vehicles.

In scenarios run for Auckland, the maximum VOC ratio for inter-regional evacuation was 92, which was established for a daytime evacuation with multiple vehicles, based on the maxima vent location and a large shadow evacuation. This suggests that the volume of traffic greatly exceeded the lane's capacity. VOC ratios were highest along motorways and in inter-regional evacuation models. In such cases, speeds were reduced to miniscule fractions of a km h^{-1}, i.e. traffic had come to a virtual standstill. The maximum VOC ratios for intra-regional evacuations however, were much lower (maximum = 6.56), a trend which was also depicted in the total network travel times (i.e. the cumulative time for all drivers to reach their final destinations). For the inter-regional evacuations modelled, clearance time was found to be in the order of days to years. Intra-regional evacuation clearance times were found to be much lower, ranging from one to nine hours. This was regardless of the level of network detail or number of connectors employed.

DISCUSSION

Impact of Diurnal Population Shifts on Evacuation Demand

Neighbourhoods with large daily population influxes are likely to be more difficult to evacuate during the day than at night because a greater demand will be placed on a static infrastructure network. Neighbourhoods with net losses will become easier to evacuate during the day (Tomsen 2010). A key finding of the population evacuation demand results is that demand values are high around the CBD during the day but become more widely dispersed at night. This can be attributed to a diurnal population shift, which is likely due to commuters moving away from the CBD toward peripheral, residential areas during the evening.

Night-time evacuation demand figures are significantly lower for the low-mobility population than the general population. In terms of geographic distribution, the North Shore has much less of a low-mobility evacuation demand as residents living there have greater access to motor vehicles. The reverse is true for South Auckland, which, when utilising the smaller two evacuation radii, nearly rivals the CBD as the peak evacuation demand centre. Daly et al. (2007) estimated that the maximum combined capacity of the Auckland public transportation system for an initial outward movement, assuming all public resources could be mustered, was just under 50,000 passenger seats. This capacity is more than sufficient for all night-time low-mobility evacuation models in our study, but not the 5 kilometre and 8 kilometre daytime low-mobility evacuation models. With a 5 kilometre evacuation radius, there is roughly a 2% chance this capacity will be exceeded during a daytime evacuation. This probability grows to roughly 35% with an 8 kilometre evacuation radius. Similar issues associated with evacuating low-mobility populations are frequently encountered worldwide. For example, during Hurricane Katrina, there were some clear failures, particularly when it came to evacuating low-mobility groups in New Orleans (Wolshon 2006).

Bridge Vulnerability and Importance of Maritime Transport

In terms of transportation networks, a worst-case scenario for Auckland is a volcanic eruption impacting both bridges connecting the isthmus in the same direction; a 1.7% and 1.5% chance for both north and both south bridges respectively, for a 5 kilometre evacuation radius. Such an event would virtually sever all motor vehicle movement into and out of Auckland in that direction. Values for public transport network accessibility for the population with no access to a private vehicle were always greater than night-time values for the entire population. This makes intuitive sense, as people without a vehicle would likely live closer to public transportation assets to fulfil any day-to-day transportation requirements. It should be noted that while motorways and public transit routes are likely to become jammed, not helped by the constraining geography in Auckland, the likelihood of such congestion on the waters of the Waitemata and Manukau Harbours is low. Boats, unlike most cars and buses, are capable of transporting extensive cargo and are often kept stocked with essential survival materials including food, water and clothes. Therefore, although non-traditional, a part-maritime based evacuation may hold several advantages in Auckland. In 2008, such an evacuation was successfully demonstrated in Chaitén, Chile where over 5,000 people were evacuated by boat shortly before the town was inundated by lahars originating from the nearby volcano (Major and Lara2013, Wilson et al. 2009). Marinas and ferry terminals are more limited in numbers than boat ramps in Auckland and the accessibility figures reflect this. Boat ramps however, have the major disadvantage of additional preparation time required as many boat owners would initially have to return home from work during the day to collect boats for an evacuation, perhaps via congested roads. Despite this, the potential for maritime evacuation by private boats and ferries should not be underestimated in Auckland.

Relative Evacuation Difficulty as Revealed by P/EC Ratios

Micro-scale vulnerability analysis using P/EC ratios can be produced anywhere that population and road network data are available. In this study, such ratios were found to be effective in determining which areas, due to their infrastructure design, pose greater difficulties for evacuation. A comparison of daytime and night-time P/EC ratios (Figure 6) shows that the number of neighbourhoods with relatively large P/EC ratios is low. However, these neighbourhoods are of greatest concern as they will likely be more difficult to evacuate on a micro-scale. Two of the five neighbourhoods with the greatest net population gain during the day, the CBD and Auckland Airport, also have the two highest P/EC ratios during the day. This indicates that neighbourhoods experiencing the largest daytime population influx are some of the most difficult to evacuate during the day because of the inflexible infrastructure setup. A similar trend is observed for neighbourhoods with high night-time population influx. Neighbourhoods such as Wattle Farm in South Auckland and Eastern Howick Beaches in East Auckland (Figure 1), with the greatest net population losses during the day, also have some of the highest night-time P/EC ratios. This is most likely attributed to the return of school children and workers during evening hours. This suggests that residential neighbourhoods that experience the greatest net population influxes at night are also some of the most difficult to evacuate at night as a result of infrastructure setup. Overall, however, the highest P/EC ratios during the day are far higher than the highest P/EC ratios during the night. This indicates that evacuations in the top-five most challenging neighbourhoods are much more challenging during the day than at night. This can be attributed to the larger number of residential neighbourhoods than business and industrial neighbourhoods in Auckland. This reflects a metropolitan standard; in order to support an urban core, multiple feeder (i.e. residential) neighbourhoods are required.

Vulnerability trends related to boundary types also emerged during the study. Significantly higher P/EC ratios were experienced by neighbourhoods with both water and motorway boundaries than neighbourhoods with neither. Contrasting P/EC ratios observed for neighbourhoods with the different boundary types (i.e. high P/EC ratios for those with motorways and low P/EC ratios for those with water during the day) are likely attributable to the neighbourhood type. Business and industrial neighbourhoods are more likely to be located near motorways to enable easy access for workers and goods. Residential neighbourhoods, in contrast, are more likely to be located close to water features as they provide aesthetic and recreational value. Motorways offer neither of these. The minima Car/EC ratios at or bordering zero for some neighbourhoods may be due to the fact that some areas, such as Druces West and Sylvia Park (both in South Auckland), have virtually no residential population and are primarily used for business purposes. Such areas will likely be easy to evacuate at night but will be difficult to evacuate during business hours when there are high populations and limited capacities.

Intra-regional vs. Inter-regional Evacuation

The preference for Auckland Civil Defence and Emergency Management is to localise evacuations as much as possible, moving evacuees outside the danger zone (via a welfare facility for registration purposes) but keeping them close to home and within the region (MCDEM 2008). There were 151 Civil Defence and Emergency Management designated welfare facilities at the time of this study, including schools, churches and sports stadiums. More than half of Auckland's population were located within 5 minutes driving distance of the nearest facility and virtually all Aucklanders can access one within a 15-minute drive. However, the nearest welfare facility will not always be available or suitable for evacuation purposes (for example it might lie in the evacuation zone or be impacted by localised critical infrastructure disruption). In such cases it will be necessary for evacuees to travel greater distances until they reach an alternative welfare facility in a safe location before heading to their final destination, wherever that may be. During a large phreatomagmatic eruption, critical infrastructure across Auckland may be crippled and an inter-regional evacuation may be necessary. Prior to this study, the best estimate of an evacuation clearance time for Auckland was developed during Exercise Ruaumoko in 2008. During this exercise, civil defence authorities allowed less than 24 hours in order to evacuate a 5 kilometre radius zone (Horrocks 2008a). However, this clearance time was only postulation and lacked any scientific backing. Though a conclusive figure will be unavailable until an actual evacuation is conducted, the next best solution, deriving scientifically backed estimates by means of evacuation modelling, was undertaken in this study.

Marco-scale evacuation vulnerability analysis showed that clearance time for inter-regional evacuation was in the order of days to years. This indicates that congestion is not limited to specific links, but rather endemic in the network due to limited capacities and excessive demand. Evacuees would most likely walk rather than wait in traffic congestion for such prolonged periods and the capabilities of the TransCAD model can therefore be deemed somewhat limited for computing movement times over such large distances especially with heavy congestion. In essence however, it can be concluded that the Auckland road network is ill suited for large-scale egress movements at present. Despite this, one major trend was identified from the inter-regional modelling: as more detail is added to the road network, thus allowing evacuees more route choice flexibility, clearance times are reduced. In the future it would be worthwhile recalculating macro-scale evacuation vulnerability using an evacuation modelling program specifically designed to accommodate high congestion levels, which could also be used to validate intra-regional evacuation results.

The relatively short clearance times for intra-regional evacuations mean that there is considerable merit to intra-regional over inter-regional evacuation from a mobility standpoint. The reasons for lower intra-regional clearance times are heavily associated with path lengths. While inter-regional evacuees commonly traversed roughly 150 kilometres of road network, intra-regional evacuees seldom travelled further than 10 kilometres. Network capacity is another contributing factor. VOC analysis confirmed that motorways would be congested in all evacuation scenarios, regardless of the amount of network detail. Yet, intra-regional evacuations are less dependent on the motorway network than inter-regional evacuations, as there are more alternate routes due to dense urban development in the city. It should be noted that all of the clearance time figures represent the movement times once evacuees have reached the Traffic Analysis Zone boundary and do not include preparation times (estimated to be 1 to 2 days in Auckland (Auckland CDEM Group 2008b)) or intra-zonal movement times.

CONCLUSIONS

This study has numerically demonstrated a future eruption in the AVF will almost certainly create considerable evacuation demands, which require considerable planning. Although the locations of new eruptions are unknown, adopting a non-specific eruption vent approach has allowed the impacts at all locations, and population and infrastructure vulnerability to be modelled and assessed. Our key conclusions and recommendations are:

- Eruptions in the central Auckland isthmus area are likely to generate high evacuation demand, peaking at just over half a million evacuees in places. This is particularly relevant given that rapid evacuation is likely to be needed, i.e. within 48 hours. High P/EC ratios in the same regions indicate that transportation infrastructure may struggle to cope. Based on a 5 kilometre evacuation radius, there is just a 4% chance that an eruption would generate no evacuation demand.
- Daytime eruptions near Auckland Airport and in business districts, particularly the CBD, will generate high population evacuation demand. Daytime maximum population evacuation demand figures are around three times higher than night-time eruptions in these areas due to the large population influx from residential suburbs during the day. There is also a substantial low-mobility population in the central Auckland area during the day, largely attributable to the number of students. Indeed the predicted daytime maximum low-mobility evacuation demand may be more than twice the seating capacity of Auckland's total public transportation assets meaning that return trips may be required to accommodate such demand. Large daytime increases in population evacuation demand combined with high P/EC ratios may favour an evacuation at night in some areas (such as the CBD) to avoid severe congestion (although this needs to be counterbalanced by the limited visibility and other constraints of evacuation by night). Delaying the call to evacuate until the majority of people are home, or advising people to remain at

home to await an evacuation decision, should be considered. This supports the consideration in the current mass evacuation plan for Auckland which favours evacuation from home due to likely reduced preparation time (Auckland CDEM Group 2008b).

- The bus network is accessible by foot to nearly the entire Auckland population but, as with private transport, is subject to road congestion that commonly plagues evacuations. There is a moderate chance (13-39% depending on the eruptive magnitude) that one of the four motorway bridges would be impacted and an up to 8% chance that multiple bridges facilitating evacuation in the same direction are impacted. Any impacts to bridges will likely increase congestion and further disrupt evacuation. Transportation by boat and rail is significantly less accessible but these modes rarely experience major congestion in Auckland. For this reason, the value of maritime evacuation by ferries, and private boats using boat ramps and marinas, should not be underestimated. This could be particularly beneficial in neighbourhoods with high population/exit capacity that have both water and motorway boundaries. Maritime evacuation would play to Auckland's geographic strength: its double harbour access, and would likely ease congestion on available motorways. In addition, the Ports of Auckland east coast seaport (adjacent to the CBD) provides an opportunity for people to be evacuated by a non-traditional mode of maritime transport, namely cargo vessels.
- Each of the modelled TransCAD scenarios experienced massive congestion, as thousands of evacuees flooded a limited number of evacuation routes. Congestion was not limited to specific links, but rather endemic in the network due to limited capacities and excessive demand. Intra-regional, rather than inter-regional, evacuations were favoured, reducing total network clearance times by multiple orders of magnitude (from days to between one and nine hours). This suggests that evacuation to destinations within the Auckland region should be given preference over evacuation to destinations outside the region if possible.
- TransCAD results also suggest that route choice flexibility will likely reduce the clearance time for evacuation. Such choices may be critical for inter-regional evacuations to be completed within manageable timeframes. However for this to work, evacuees must either have prior knowledge of the various network

connections or be informed of them using signage and/or other communication methods.

- The survivability of Auckland's critical infrastructure during volcanic activity is an important unknown. Functional transportation networks are fundamental for evacuation purposes. Whether infrastructure assets, including bridges, tunnels, traffic lights, signs and motorways will be functional following volcanic activity such as tremors and ash fall determines which routes are available to evacuees and radically impacts egress movements. Further study is required in this field.

AUTHORS' CONTRIBUTION

Supported by the expertise and direction of JML and MG, ET acquired data, planned and conducted the research and undertook all statistical analysis for the study. TMW contributed knowledge and skills, particularly regarding the modelling component of the work and DMB participated through the writing and editing of this manuscript. All authors have read, reviewed and approved the final manuscript.

ACKNOWLEDGEMENTS

We are grateful to the Rotary Foundation for providing funding for ET to conduct the research. We thank Statistics New Zealand, the Ministry of Education and the former Auckland Regional Council for the supply of data. We also thank Associate Professor André Dantas, in Civil Engineering at the University of Canterbury at the time of the research, for his support with the TransCAD studies. JML gratefully acknowledges support from the New Zealand Earthquake Commission.

REFERENCES

1. Allen SR, Smith IEM (1994) Eruption styles and volcanic hazard in the Auckland Volcanic Field, New Zealand. Geoscience Reports of Shizuoka University 20:5-14

2. ArcGIS (2009) GIS software: version 9.3.1. Environmental Systems Research Institute (ESRI), Redlands, California.
3. Auckland CDEM (2013) Auckland Volcanic Field contingency plan. Available from: http://www.aucklandcivildefence.org.nz/About-Us/Document-Library/Supporting-Plans/ . Accessed 22 June 2013
4. Auckland CDEM Group (2008a) Final exercise report: exercise Ruaumoko '08. Auckland Regional Council, Auckland, New Zealand.
5. Auckland CDEM Group (2008b) Mass evacuation plan: group functional plan P2. Auckland Regional Council, Auckland, New Zealand.
6. Auckland Regional Council (2009a) Monitor Auckland: tracking our progress. Auckland Regional Council, Auckland, New Zealand.
7. Auckland Regional Council (2009b) Trends and issues (transport challenges): regional land transport strategy working report. Auckland Regional Council, Auckland, New Zealand.
8. Baker EJ (1991) Hurricane evacuation behavior. International Journal of Mass Emergencies and Disasters 9:287-310
9. Bascetta C (2006) Disaster preparedness preliminary observations on the education of vulnerable populations due to hurricanes and other disasters testimony. In: United States Government Accountability Office (ed) Testimony before the special committee on aging, U.S. Senate, Washington D.C.
10. Bebbington MS, Cronin SJ (2011) Spatio-temporal hazard estimation in the Auckland volcanic field, New Zealand, with a new event-order model. Bull Volcanol 73(1):55-72
11. Beca Carter Hollings and Ferner Ltd (2002) Contingency plan for the Auckland Volcanic Field. Auckland, New Zealand. http://www.arc.govt.nz/albany/fms/main/Documents/Council/Civil%20defence%20emergency%20management/Publications%20&%20factsheets/Auckland%20Volcanic%20Plan%20(VCP).pdf . Accessed 16 June 2013
12. Belousov A, Voight B, Belousova M (2007) Directed blasts and blast-generated pyroclastic density currents: a comparison of the Bezymianny 1956, Mount St Helens 1980, and Soufrière Hills,

Montserrat 1997 eruptions and deposits. Bull Volcanol 69(7):701-740 10.1007/s00445-006-0109-y

13. Blake S, Smith I, Wilson C, Leonard G (2006) Lead times and precursors of eruptions in the Auckland Volcanic Field, New Zealand: indications from historical analogues and theoretical modelling. Consultancy Report 2006/34, GNS Science, Lower Hutt, New Zealand.
14. Brisbane City Council (2007) Brisbane Strategic Transport Model (BSTM) version 6.0. Brisbane City Council, Brisbane, Australia.
15. Browne FL (1958) Theories of the combustion of wood and its control: a survey of the literature. Forest Products Laboratory, U.S. Forest Service Research Legacy, Report 2136
16. Chakraborty J, Tobin GA, Montz BE (2005) Population evacuation: assessing spatial variability in geophysical risk and social vulnerability to natural hazards. Natural Hazards Review 6:23-33
17. Church RL, Cova TJ (2000) Mapping evacuation risk on transportation networks using a spatial optimization model. Transportation Research Part C: Emerging Technologies 8:321-336
18. Cola RM (1996) Responses of Pampanga households to lahar warnings: lessons from two villages in the Pasig-Patrero River watershed. In: Newhall CG, Punongbayan RS (eds) Fire and Mud, Philippine Institute of Volcanology and Seismology and University of Washington Press, Quezon City and Seattle.
19. Cole J, Blumenthal E (2004) Evacuate: what an evacuation order given because of a pending volcanic eruption could mean to residents of the Bay of Plenty. Tephra 21:46-52
20. Cova TJ (1999) GIS in emergency management. Geographical Information Systems: Principles, Techniques, Applications and Management. John Wiley and Sons, New York.
21. Cova TJ, Church RL (1997) Modelling community evacuation vulnerability using GIS. Int J Geogr Inf Sci 11:763-784
22. Cutter CL (2003) GI science, disasters, and emergency management. Trans GIS 7:439-446

23. Daly M, Hoskin K, Leonard G (2007) Physical evacuation workshop, 2007 (facilitator comments). Auckland Regional Council, Auckland, New Zealand.

24. Dosa DM, Grossman N, Wetle T, Mor V (2007) To evacuate or not to evacuate: lessons learned from Louisianna nursing home administrators following Hurricanes Katrina and Rita. Journal of the American Medical Directors Association 8:142-149

25. Dow K, Cutter SL (2002) Emerging hurricane evacuation issues: Hurricane Floyd and South Carolina. Natural Hazards Review 3:267-285

26. Drabek TE (1986) Human system responses to disaster. Springer, New York.

27. Franzese O, Liu C (2008) Emergency evacuations, transportation networks. In: Shekhar S, Xiong H (eds) Encyclopaedia of GIS, Springer, New York.

28. Horrocks J (2008) Evacuation progression script summary, Exercise Ruaumoko 2008. Ministry of Civil Defence and Emergency Management, New Zealand.

29. Horrocks J (2008) Learning from Exercise Ruaumoko, Exercise Ruaumoko 2008. Ministry of Civil Defence and Emergency Management, New Zealand.

30. Hushon J, Kelly RB, Rubin C (1989) Identification and analysis of factors affecting emergency evacuations. Nuclear Management and Resources Council Inc. (NUMARC), Washington D.C.

31. Ketteridge AM, Clarke L, Fordham M (1996) EUROflood: evacuation 1996, technical annex 14 for the evacuation module. Flood Hazard Research Centre, University of Middlesex, Enfield, United Kingdom.

32. Klepeis NE, Nelson WC, Ott WR, Robinson JP, Tsang AM, Switzer P, Behar JV, Hern SC, Engelmann WH (2001) The National Human Activity Pattern Survey (NHAPS): a resource for assessing exposure to environmental pollutants. J Expo Anal Environ Epidemiol 11:231-252

33. Le Corvec N, Bebbington MS, Lindsay JM, McGee LE (2013a) Age, distance and geochemical evolution within a monogenetic volcanic field: analyzing patterns in the Auckland Volcanic Field

eruption sequence. Geochemistry, Geophysics, Geosystems 14(9):3648-3665

34. Le Corvec N, Spörli KB, Rowland J, Lindsay J (2013) Spatial distribution and alignments of volcanic centers: clues to the formation of monogenetic volcanic fields. Earth Sci Rev 124:96-114
35. Leonard R (1985) Mass evacuation is disasters. J Emerg Med 2:279-286
36. Lindell MK, Perry RW (1992) Behavioral foundations of community emergency planning. Hemisphere Publishing, Washington D.C.
37. Lindell MK, Prater CS (2006) Evacuation from Hurricane Katrina: Jefforson and St. Charles parishes. National Hurricane Conference, Orlando, Florida.
38. Lindell MK, Prater CS (2007) Critical behavioural assumptions in evacuation time estimate analysis for private vehicles: examples from hurricane research and planning. Journal of Urban Planning and Development 133:18-29
39. Lindell MK, Lu J-C, Prater CS (2005) Household decision making and evacuation in response to Hurricane Lili. Natural Hazards Review 6:171-179
40. Lindsay JM (2010) Volcanoes in the big smoke: a review of hazard and risk in the Auckland Volcanic Field. Institute of Earth Science and Engineering and School of Environment. University of Auckland, Auckland, New Zealand.
41. Lindsay JM, Marzocchi W, Jolly G, Constantinescu R, Selva J, Sandri L (2010) Towards real-time eruption forecasting in the Auckland volcanic field: application of BET_EF during the New Zealand national disaster Exercise 'Ruaumoko'. Bull Volcanol 72:185-204
42. Lindsay JM, Leonard GS, Smid ER, Hayward BW (2011) Age of the Auckland Volcanic Field: a review of existing data. N Z J Geol Geophys 54(4):379-401
43. Major JJ, Lara LE (2013) Overview of Chaitén volcano. Andean Geology 40(2):196-215
44. Marrero JM, Garcia A, Llinares A, Rodriguez-Losada JA, Ortiz R (2010) The Variable Scale Evacuation Model (VSEM): a new

tool for simulating massive evacuation processes during volcanic crises. Natural Hazards and Earth Systems Sciences 10:747-760

45. Marzocchi W, Woo G (2007) Probabilistic eruption forecasting and the call for an evacuation. Geophysical Research Letters 34:1-4
46. MCDEM (2008) Mass evacuation planning: director's guidelines for civil defence emergency management (CDEM) groups, DGL 07/08. Ministry of Civil Defence and Emergency Management. http://www.civildefence.govt.nz/memwebsite.nsf/Files/Director_Guidelines/$file/Mass-evacu-planningDGL07-08.pdf . Accessed 12 May 2013
47. Mileti DS, Bolton PA, Fernandez G, Updike RG (1991) The eruption of Nevado Del Ruiz volcano Colombia, South America, November 13, 1985. Commission on Engineering and Technical Systems (National Academy Press), Washington, D.C.
48. Moriarty K, Ni D, Collura J (2007) Modelling traffic flow under emergency evacuation situations: current practices and future directions. 86th Transportation Research Board Annual Meeting. Transportation Research Board, Washington, D.C.
49. Morrow BH (1999) Identifying and mapping community vulnerability. Disasters 23:1-18
50. Needham AJ, Lindsay JM, Smith IEM, Augustinus P, Shane PA (2011) Sequential eruption of alkaline and sub-alkaline magmas from a small monogenetic volcano in the Auckland Volcanic Field, New Zealand. J Volcanol Geotherm Res 201:126-142
51. New Zealand Transport Agency (2009) Motor vehicle registration statistics 2008. MVR Statistics, Wellington, New Zealand.
52. NZ Open GPS Maps (2009) Improved NZ road centrelines. Koordinates. https://koordinates.com/layer/183-improved-nz-road-centrelines-august-2011/ Accessed 10 November 2009
53. Perry RW, Lindell MK (2003) Preparedness for emergency response: guidelines for the emergency planning process. Disasters 27(4):336-350
54. Prater C, Wenger D, Grady K (2000) Hurricane Bret post storm assessment: a review of the utilisation of hurricane evacuation studies and information dissemination. College Station, TX, Hazard Reduction and Recovery Centre, A&M University, Texas.

55. Quarantelli EL (1980) Evacuation behaviour and problems: findings and implications from the research literature. Ohio State University Disaster Research Centre, Columbus, Ohio.
56. Quarantelli EL (1985) Social support systems: some behavioural patterns in the context of mass evacuation activities. Disasters and Mental Health: Selected Contemporary Perspectives 178:122-136
57. Sandri L, Jolly G, Lindsay J, Howe T, Marzocchi W (2012) Combining probabilistic hazard assessment with cost-benefit analysis to support decision making in a volcanic crisis from the Auckland volcanic field, New Zealand. Bull Volcanol 74:705-723
58. Shaluf IM (2008) Technological disaster stages and management. Disaster Prevention and Management 17:114-126
59. Shane P, Gehrels M, Zawalna-Geer A, Augustinus P, Lindsay J, Chaillou I (2013) Longevity of a small shield volcano revealed by crypto-tephra studies (Rangitoto volcano, New Zealand): change in eruptive behaviour of a basaltic field. J Volcanol Geotherm Res 257:174-183
60. Sherburn S, Scott B, Olsen J, Miller C (2007) Monitoring seismic precursors to an eruption from the Auckland Volcanic Field, New Zealand. N Z J Geol Geophys 50:1-11
61. Shulman HML (2008) Estimating evacuation vulnerability of urban transportation systems using GIS. Department of Geography, Queen's University, Kingston, Canada.
62. Sparks RSJ (2003) Forecasting volcanic eruptions. Earth Planet Sci Lett 210:1-15
63. Spörli KB, Eastwood VR (1997) Elliptical boundary of an intraplate volcanic field, Auckland, New Zealand. J Volcanol Geotherm Res 79(3–4):169-179
64. Statistics New Zealand (2006) Census of population and dwellings. Wellington, New Zealand.
65. Statistics New Zealand (2008) Business demographic statistics Table 11: business demography tables. Wellington, New Zealand.
66. Tierney KJ, Lindell MK, Perry RW (2001) Facing the unexpected: disaster preparedness and response in the United States. Joseph Henry, Washington, D.C.

67. Tobin GA, Whiteford LM (2002) Community resilience and volcano hazard: the eruption of Tungurahua and evacuation of the Faldas in Ecuador. Disasters 26(1):28-48
68. Tomsen E (2010) GIS-based mass evacuation planning for the Auckland volcanic field. Masters Dissertation, School of Environment, University of Auckland, Auckland, New Zealand.
69. U.S. Department of Transportation (2006) Catastrophic hurricane evacuation plan evaluation: a report to congress (in corporation with the U.S. Department of Homeland Security). http://www.fhwa.dot.gov/reports/hurricanevacuation/rtc_chep_eval.pdf . Accessed 01 August 2013
70. Urbanik T (2000) Evacuation time estimates for nuclear power plants. J Hazard Mater 75:165-180
71. Wilson T, Leonard G, Stewart C, Johnston D (2009) Rhyolitic eruption of Vulcan Chaitén, Chile. HAZM 401 – Geological Uncertainty, University of Canterbury, Christchurch, New Zealand.
72. Wolshon B (2002) Planning for the evacuation of New Orleans. Institute of Transportation Engineers 72(2):44-49
73. Wolshon B (2006) The aftermath of Katrina. The Bridge, National Academy of Engineering, Washington D.C. 36(1):27-34
74. Wolshon B (2008) Empirical characterisation of mass evacuation traffic flow. J Transp Res Board 2041:38-48
75. Wolshon B, Urbina E, Levitan M (2001) National review of hurricane evacuation plans and policies. Louisiana State University Hurricane Centre, Baton Rouge, LA.
76. Woo G, Grossi P (2009) A new era of volcano risk management. Risk Management Solutions Inc, Stanford, CA. https://support.rms.com/publications/Volcano_Risk_Management.pdf . Accessed 20 June 2013

Chapter 6

An Activity Theory Analysis of Boundary Objects in Cross-Border Information Systems Development for Disaster Management

Nitesh Bharosa[1], JinKyu Lee[2], Marijn Janssen[3], and H Raghav Rao[4, 5]

[1]Delft University of Technology, Jaffalaan 5, 2628BX, Delft, The Netherlands

[2]Spears School of Business, Oklahoma State University, 317 North Hall, 700 N. Greenwood Ave, Tulsa, OK, 74106, USA

[3]Delft University of Technology, Jaffalaan 5, 2628BX, Delft, The Netherlands

[4]Management Science and Systems, University at Buffalo, 325C Jacobs Mgmt Center, Amherst, NY, 14260, USA

[5]Department of GSM, Sogang University, Seoul, Korea

ABSTRACT

One of the main challenges in cross-border disaster management is the development and use of information systems that cater the needs of heterogeneous relief agencies, policies, activities and cultures. Drawing upon activity theory, this paper examines cross-border information systems development for disaster management. We infuse the concept of boundary objects into activity theory by the characterization of the artifacts. This allows articulating how the socio-technical objects are meshed with the process of cross-border collaboration for systems development. Our longitudinal ethnographic field study on a cross-border flood management project, VIKING, revealed how the project was empowered and developed by four key boundary objects, i.e. the governance structure of the program, two information systems (a disaster management information system and an online collaboration portal), and recurring cross-border exercises as an evaluation and feedback mechanism. The selective institutionalizations of these key boundary objects helped the participants overcome various contradictions existed in the systems development. The study results also show that both goal-oriented actions and boundary objects can affect the outcomes of long-term large-scale disaster management systems development.

BACKGROUND

Disaster management (DM) entails a range of complex interdependent activities involving many collaborating organizations. It is particularly so if the disaster affects multiple countries as we have seen in the 2004 Indian Ocean tsunami or multiple regions in a single country as has been seen in the floods of Oct-November 2011 in Thailand. Although information systems are often adopted to expand the scope of management in both private and public sectors, information system development (ISD) is by itself a complicated endeavor, especially in inter-organizational and international settings where organizations with different goals, existing technologies and cultural backgrounds have to collaborate [1-3]. DM organizations (DMOs) have attempted to leverage information systems (IS) in an effort to effectively coordinate various disaster management efforts. Developing an IS to support complex cross-border DM activities, however, is an overwhelming task even for most capable government authorities for a number of reasons.

First, the development of such a disaster management information system (DMIS) requires cross-organizational collaboration between multiple autonomous organizations. These organizations often have incompatible responsibilities, procedures and objects [4], highlighting the diverse requirements for cross-organizational information sharing systems that must accommodate a wide range of information needs and flows [5]. Second, the relational ties between many DMOs are often temporary and weak as inter-organizational collaboration is not necessary in their day-to-day operations during non-disaster periods [5]. Most DMOs rarely need cross-organizational information sharing, and thus it is hard to keep them interested and committed to the development of a DMIS for occasional use. The special systems requirements for cross-organizational and cross-border DMIS, such as reliability, interoperability, functionality, and accessibility, present yet another reason that makes it a daunting task. Accordingly, we conceptualize a cross-border DM ISD as a complex collaborative process in which many stakeholders conduct a wide range of activities centered around various socio-technical artifact (e.g., inter-organizational structure, shared systems development environment, communication channels, etc.), some of which may change as the development process evolves.

Due to the importance of effective disaster management, there are an increasing number of scholars picking up on the challenges of cross-organizational DM ISD e.g., [6-8]. While some contributions have been made on understanding cross-organizational ISD [9], studies on cross-border collaboration for DM ISD are scarce, with only a few exceptions e.g., [1, 10]. This research aims to understand the nature of boundary objects that facilitate long term cross-border DM ISD process. For this purpose, a collaborative DM ISD project between Germany and the Netherlands, entitled Program VIKING, is investigated. Employing an ethnographic field study approach, this study identifies the collaborative activities and boundary objects that have made substantial positive effects on the cross-border DM ISD process. The field study was conducted using three data collection techniques: (1) semi-structured, face-to-face interviews, (2) participatory observations, and (3) document analysis. We draw on activity theory (AT) to frame the interactions among the activities, objects, and advances in the DM ISD.

In the next section, we discuss about activity theory and boundary objects that provide a theoretical foundation for our study of cross-border DM collaborations. Then we present our research design and methodology along with a brief introduction to the studied cross-border DM collaboration case. Next, the boundary objects that play a pivotal role in the success of the collaboration activity are identified. We conclude the paper with key research findings and discussions on the theoretical and practical implications of the findings.

Theoretical Foundation

Activity Theory

To understand the critical success factors for a cross-border ISD, the collaboration activities must be systematically specified, and the mechanisms through which the critical factors influence the collaboration must be articulated. Activity Theory (AT) [11,12] fits well with our research objectives. AT portrays a social system as a group of individuals engaging and interacting with their environment [13], which will result in output objects (a.k.a. artifacts) [14,15]. The number of interactions and resulting artifacts can increase as the social system persists over time, while some of the artifacts can affect the future activities and structure of the system. AT underlines the role of emergent artifacts in relation to the activity system and can untangle human-object relationships to observe the unfolding of the activity system (e.g., cross-border DM ISD project group) over time. As such, AT provides us with a useful analytical framework to examine the activities and artifacts of a cross-border DM collaboration system that can facilitate and prolong the system.

AT defines a social system as an activity system that consists of various components, including the subject (i.e., agent or actors), objectives (i.e., purposes of actions), tools, rules, community, and division of labor [11]. A subject is a person or a group of people engaged in an activity. Acting as subjects, actors can take a set of collated or extended actions in order to achieve a goal (i.e., a desirable outcome at the individual action-level), which should collectively contribute to the system-level activity. An objective is the desired outcome [16] held by the subject and motivates the activity, giving it a specific direction

[13]. Transforming the objective (e.g., reduction of flood damage) into an outcome (e.g., cross-border DM initiative, development of DMIS, effective disaster response) motivates the existence of an activity system. There can be a multiplicity of objectives that may be revised or evolve over time in a long-term DM project. Community refers to the group of subjects that share the objectives of an activity system. In the context of cross-border ISD collaboration for DM, a large number of DMOs and IS vendors/consultants from the participating countries will comprise the community of the activity system. The relationships between subject, objectives and community are mediated by tools, rules and division of labor. An activity achieves an objective via a development process that typically has multiple steps or phases [17]. A "tool" can be anything used in the development process, including both material tools (e.g., servers, workstations) and tools for thinking (e.g., online forums, pressure); "rules" cover both explicit and implicit norms, conventions and social relations within the community, while "division of labor" refers to the explicit and implicit organization of the community of an activity system as the subjects are involved in the transformation process [17]. In the cross-border DM context, the rules and division of labor include institutionalized business processes and (inter-)organizational structures (e.g., national incident management system) that can widely vary from one country to another as well as among different types of collaborating organizations (e.g., firefighters vs. law enforcement vs. domain experts/hazmat).

AT was initially adopted in the IS field to understand human-computer interaction [13,16], and has also proved to be valuable in analyzing complex processes such as cross-organizational data model development i.e., [18,19]. Shankar et al. [20] outline five fundamental principles of AT based on [11].

- The prime unit of analysis is an activity system, where an activity system refers to a collective, artifact-mediated, and object-oriented system of actions, which exists in a network of related activity systems.
- Every activity system is subject to multiple perspectives and beliefs (i.e., multi-voicedness). The division of labor in an activity system induces diverse positions for the diverse groups of actors involved in the activity. The division of labor is a primary component of any activity system, impressed with the rules and conventions.

- The issues and capabilities of an activity system can only be analyzed by their own history (i.e., historicity). Thus, AT implicitly incorporates the structurational view of organizational transformation where human actions and institutional structures can mutually influence each other [21-23].
- Contradictions are the sources of changes and developments. Contradictions amass structural tensions within and between activity systems.
- An activity system can expand its capability (i.e., expansive transformations). Expansive transformation refers to a process through which an activity system expands its capability by redesigning its own structure and changing its actions in order to resolve a challenging contradiction. As suggested in the previous (4th) principle, activity systems generally enter lengthy phases of fundamental transformations when a contradiction cannot be resolved by existing structure and practice. Such fundamental transformation, if successful, will results in a reoriented activity system [24] with an expanded capability required in the new environment.

The principles stated above make AT appropriate for examining cross-border DM ISD characterized by collective actions, historicity, contradictions and expansive transformations.

Boundary Objects

AT recognizes the influences of both existing structure (i.e., rules, community, and division of labor) and human agents (e.g., human subjects using instruments or initiating an expansive transformation) [11]. Furthermore, AT appreciates shared understanding (e.g., expansive learning by the community members) as a means to improve social structures [25,26]. In relation to shared understanding, it is essential to understand the concept and role of boundary objects in cross-border collaboration activities.

Boundary objects are referred to as conceptual or physical artifacts that reside in the interfaces among organizations [27]. From a functional perspective, a boundary object is an artifact shared by a community of subjects (e.g., DMOs in cross-border collaboration) that work together to reach their individual goals. Such an object can be a material thing, but it can also be less tangible or totally intangible (e.g., plans, common ideas) as long as it can be shared for manipulation and transformation by the participants of the activity. For an object to serve as a boundary object, it must be "both plastic enough to morph to local needs and constraints of the several parties employing them, yet robust enough to maintain a common identity across sites" p. 393 [28]. Boundary objects help interacting organizations facilitate cross-organizational communication and form an organizational identity [27], while they can also act as gatekeepers that selectively filter information between the organizations. Hussenot and Missionier [29] depict them as bearers of compromises that promote cooperation between the stakeholders (p.274). Therefore, boundary objects should be able to bear various meanings assigned by different organizations while serving as a common reference point to the members of multiple organizations when they engage in mutual practice [28].

IS researchers have recognized the potential of IS as a boundary object that can facilitate boundary spanning [30,31]. Accordingly, boundary objects may include physical product prototypes, design drawings, shared IT applications, standard business forms, or even shared abstract constructs such as product yield [32].

For a boundary object to emerge, a new joint field of practice must be produced [25]. For example, a shared information system can be a joint field of practice. However, not every IS becomes a boundary object in reality because human agents in some organizations may not see its local usefulness or the IS may fail to establish a common identity across all organizations [25]. The VIKING alliance was originally an outcome object (a.k.a. artifact), but it became a part of the structure and influenced the transformation process of ISD. Hence it served as a boundary object that developed a shared identity and functioned as a communication hub.

Research Design & Methodology

This study adopts a qualitative, ethnographic field study approach in order to identify important factors (e.g., conditions and actions) for successful development of cross-border DM ISD. We chose a qualitative research methodology to develop a comprehensive understanding from rich field data. Qualitative methodologies are more appropriate where the research question is exploratory in nature and exists within a broader sociological context, thus necessitates rich descriptions of the social environment [33]. Furthermore, such an approach allows the exploration of unforeseen relationships and offers better insights into the inter-dependencies among the factors captured in the study [34]. Research on the evolution of a partnership, the controversies and compromises between organizations across a national border requires data that is both rich in contextual information and deep in understanding [35]. The field study was conducted over a 9-month period, covering over 4 years of cross-border collaboration that developed a DMIS at the Dutch-German border, dubbed Program VIKING. This program is described in a later section in details. Both the process of the DM ISD and the use of the DMIS in a cross-border exercise were investigated.

This field study has employed three data collection techniques: (1) semi-structured, face-to-face interviews, (2) participatory observations, and (3) document analysis. A series of semi-structured face-to-face interviews [36] were conducted with representatives of the first responder and water management agencies, the lead IT developers, and technical and managerial staff of Program VIKING. The interview protocol contained questions concerning the project initiation (e.g., history, scope, leadership), the technological solutions developed and their development process (e.g., participants, negotiation, objectives, conflicts), and factors affecting the success of the project. In the formal interviews, subjects were asked a series of open-ended and unstructured questions, augmented by follow-up questions for clarification. This procedure allowed the respondents to elaborate on the issues and ask for clarifications from the interviewees to ensure a better understanding of the context and concepts in the study [37]. Informants also discussed their motivations for participation in the program, key milestones, and technical and non-technical challenges that they had encountered thus far. Formal interviews were conducted during a cross-border DM exercise (a.k.a. ROAR). Twelve formal interviews were completed with nine key informants; three of them were interviewed twice. All formal interviews were recorded and transcribed for data analyses.

In addition to the formal interviews, we observed the use of the DMISs developed by the VIKING alliance during the three-day ROAR exercise. Participatory observations allows researchers to collect rich data in a direct way [38] while distortion of the results can be reduced to a minimum through direct interaction with research objects [39].

In total, three days of training with approximately 500 participants from both nations was observed. At least one member of the research team observed each training run, and one member attended three full days of the ROAR exercise. The data collection criteria that guided the interviews also guided the observations. In addition, the observations allowed us to conduct opportunity-based informal interviews with DMIS users. Thanks to the cooperation and support of the Program VIKING committee, we were allowed unrestricted access to the VIKING file sharing system that hosts internal documents generated during Program VIKING. We analyzed both Dutch and German documents to enrich our understanding of the program and to clarify our interpretations of the interview and observation data. Examples of documents include project meeting reports, software requirements specification and DM ISD schedules. The lists of interviewees and full document corpus are included in the Tables 1 and 2.

Table 1: The subjects of formal interviews – the primary affiliation and the role in Program VIKING

ID	Organization	Role
1	Wateboard Gelderland	Coordinator water board
2	Decis Lab	Project manager Exercise Integration
3	Geonovum	Project manager system development and maintenance
4	Province Gelderland	Program coordinator for the Netherlands
5	Fire department	(former) Chief of the regional Fire department
6	FLIWAS	Application developer
7	Justice department	Auditor
8	Police Academy	Coordinator control room
9	Busy to	Program VIKING evaluator

Bharosa *et al.*

Bharosa *et al. Security Informatics* 2012 1:15, doi:10.1186/2190-8532-1-15

Table 2: Corpus of Analyzed Documents

·	Notes of the steering committee meetings from 2004–2006
·	Annual project plans from 2004–2009
·	Internal reviews of the cross-border exercises (both in Dutch and German), including user satisfaction with the information sharing systems
·	ICT architecture planning and development documentation
·	Helpdesk log files of the ISs used during cross-border exercises
·	Observation notes of IS use during exercises
·	Cross-border exercise scenario and scripts
·	Analysis and description of the available information systems for flood management in both countries
·	Analysis and description of the necessary cross-border information flows
·	User manuals for FLIWAS
·	Maintenance reports of FLIWAS
·	Service level agreements on FLIWAS and outsourcing plans for 2009
·	The education and training program for VIKING
·	Functional requirements of the FLIWAS components

Bharosa *et al.*

Bharosa *et al. Security Informatics* 2012 1:15, doi:10.1186/2190-8532-1-15

Analytical Framework

The literature in the field of organization studies argues that organizations, collective activities, and organizational processes are all constructed through the relationship between agencies (i.e., subjects) and objects [30]. To better understand the relationships between agencies and objects, we must observe their entanglement over time. As Hussenot & Missionier [29] have demonstrated, boundary objects evolve over time. Although the transformation and nature of boundary objects have been researched in cross-organizational environments e.g., [27,40], previous contributions have not clarified the relationships among activities and boundary objects nor the implications of changes in boundary objects for the agencies that use them. This calls for more careful consideration of how the material and social objects become "entangled" in the process of collaboration. We propose an analytical framework (Figure 1) that augments previous studies by analysing boundary objects from an AT perspective.

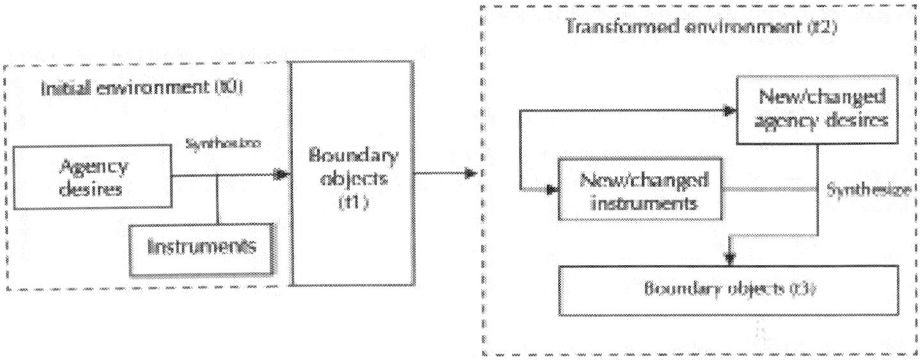

Figure 1: Analytical framework.

The proposed framework suggests that boundary objects can be synthesized by an activity system, which can in turn be morphed into the system and act as expansive solutions for subsequent cross-border DM ISD collaboration. The framework embraces AT elements such as agencies (subjects), instruments (tools), artifacts (outcomes), and environments, while environments encompasses the concepts of rules, community, and division of duties. As an activity system starts (t0 in Figure 1), it continues to exist for an extended period of time, some artifacts of its actions become boundary objects (t1). Such boundary objects transform the structure and behaviors of the activity system (t2), helping the system achieve its goals (t3). This framework emphasizes the pivotal role of boundary objects and delineates the dynamics among the structure of an activity system, the actions of its actors, and its environments.

Our framework can be directly applied to a transforming cross-boundary collaboration system (i.e., cross-border DM ISD program), captured over time. In such a system, the desires of the subject(s), instruments and actions in the given environment shape recursive transformation processes. Some, but not all, artifacts resulting from the subject's purposeful action (i.e., outcomes) at one point of time (t1) may become a part of the structure for actions at a subsequent time (t2). Such a structure would include subjects, instruments, and environments, while the environments encompass the concepts of rules, community, and division of duties. Whether an artifact becomes a part of the system structure or not depends on the attributes of the artifacts, thus we dub "a selective institutionalization process". We posit that the acceptability of these attributes to the various stakeholders in the system (i.e., community) is the key determinant of the selective institutionalization process. Artifacts that possess the characteristics of boundary objects, i.e., well accepted and shared by the stakeholders [32,41], are very likely to become a part of the activity system's structure and exert significant impacts on subsequent activities, especially when the community of the system includes a large number of stakeholders (e.g., cross-border DM ISD). Accordingly, we pay special attention to the selective institutionalization processes in Program VIKING, where some artifacts become formal boundary objects in the system. This will enable us to identify the conditions (e.g., existing rules, available instruments, imposed division of labor) and actions (i.e., creation, institutionalization, and utilization of the artifacts) that can help achieve a desirable state of the system. The focus on institutionalized artifacts will also suggest why the conditions and actions are important in view of the internal contradictions addressed via the boundary objects, and how the expansive learning and solutions that addressed the contradictions are implemented (e.g., the subject, object, instruments, action) [42]. We use the analysis framework as a guideline for our data encoding and analysis activities.

Coding scheme and Analysis Process

Collected interview data were coded using a coding scheme keyed to conventional terms used to specify the conceptual or physical objects in the program (e.g., Viking alliance, User group, Waterboard, FLIWAS, Viking Cockpit, ROAR, EU guidelines, program manager). Each instance of the objects was associated with an element of activity (i.e., actor, instrument, environment, and artifact) according to the analysis framework shown in (Figure 1)The association was recorded at instance level (i.e., each appearance in the data) because one object could be categorized into more than one element, depending on the context where the object was mentioned. For example, the Viking alliance was an artifact in the context of alliance formation activity, but it played a role of an actor, an instrument, and an environment in a later stage as it became a central part of the structure of Program VIKING.

We operationalized boundary objects as the objects mentioned by multiple data sources (i.e., interviewees from different organizations). With this operationalization, we searched for the key boundary objects that once emerged as artifacts and later appear again as another element (e.g., actors, instruments, environment), assuming that they were the (enablers of) expansive solutions that significantly facilitated the cross-border DM ISD activity. Upon the identification of the expansive solutions, other objects and theoretical antecedents of collaboration factors were related to the solutions. This analysis was to clearly understand the conditions and actions that triggered (or were triggered by) the emergence of expansive solutions. As expected, some of the identified boundary objects had a large number of connections to other coded objects, suggesting that they played critical roles in Program VIKING. (Figure 2) is a visual representation of the concept coding scheme. The results of the analysis are described in details in the research findings section. While this research is designed as a qualitative study, this coding scheme confers some level of objectivity on the analysis procedure. The results of the interview data analysis were triangulated with the researchers' insights derived from observations of the IS use, discussions with users (DMOs) during the exercises and analyses of archival data, which enabled us to build a clear, in-depth understanding of the studied case from the compiled qualitative data [43].

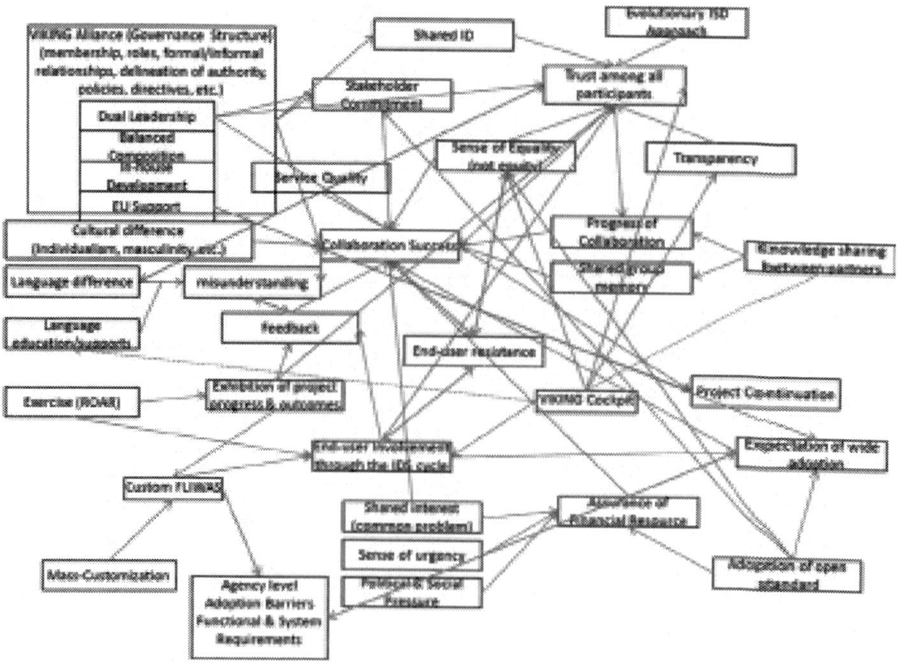

Figure 2: Concept coding scheme.

Case Background: program Viking

In 1993 and 1995, the area surrounding Rijn River across the border of the Netherlands and Germany was affected by major floods. This area belongs to the Province of Gelderland, the Netherlands and the Province of Nordrhein-Westfalen, Germany. Both sides of the border together, this area has over 1.7 million inhabitants spread over 56 municipalities. During the floods, cattle and over 250,000 people had to be evacuated for 5–20 days. Fortunately, the dikes did hold the floods, but if they were destroyed, some parts of the area would have been submerged under 5 meters of water. This area is particularly prone to major floods as the Rijn River flows through Germany into the Netherlands before it meets the North Sea. A flood of the Rijn in one area may result in a flooding in another area as far as 40 kilometers away, which will not stop at the countries' border. In preparation for major floods, the local and regional governments in both countries have heightened dikes and stimulated development of information systems for flood management. After a myriad of agency specific, local and regional projects, an inspection of safety and flood management in the area called for integration and collaboration between Dutch and German counterparts in both water and disaster management. This led to the instantiation of Program VIKING. Program VIKING focuses on improving four aspects of flood and disaster management capability: (1) the operational processes, (2) the information sharing systems (architecture and maintenance), (3) cross-border collaboration between DMOs, and (4) the education and training of multi-disciplinary DM teams.

There are some noticeable differences between the DM environments of the two countries (see Table 3). The differences can be understood on different levels, including the cultural, technological, and institutional levels. On a cultural level Hofstede's [44] study revealed that the Germans are generally more formal than the Dutch in their ways of interactions. On the technological level we found that the two countries use totally different sets of information systems for flood management. On the institutional level, the main difference between Germany and the Netherlands is the division of duties and jurisdictions. In Germany, municipalities have their own police, fire brigade and ambulance services, whereas in the Netherlands these DMOs are organized around 25 safety regions, each of which covers 10–20 municipalities.

Table 3: Differences between two nations

Level	Aspect	The Netherlands	Germany
Cultural	Main language	Dutch	German
	Cultural differences	Small power distance, informal, flexible, modest, individualistic	Big power distance, formal, inflexible, strict, collectivistic-committed to institutional structure
Technological	Flood information systems	1. PoldEvac,	1. HZG,
		2. Hoogwater InformatieSysteem (HIS),	2. X-border GDI,
		3. Geautomatiseerd Draaiboek Hoogwater (GDH),	3. Deichinformationssystem,
		5. POIRE,	4. HOWISS,
		6. SHERPA	5. COBRA,
			6. DISMA

Institutional	Institutional structure	Decentralized relief agencies with high autonomy	Centralized relief agencies, moderate autonomy
	Government standards for IS development	NORA	INSPIRE
	Disaster classification and coordination structure	GRIP structure containing four levels, flood is immediately GRIP 4 (highest level)	Stuffe structure (1,2,3)

Bharosa *et al.*

Bharosa *et al. Security Informatics* 2012 1:15, doi:10.1186/2190-8532-1-15

As a result, various DMOs are coordinated by the mayor of the largest municipality in the safety region in the Netherlands, whereas in Germany the coordination is distributed over three levels: districts, regions and states. The province of Gelderland and the Regional Government (Bezirksregiering) of Düsseldorf have no direct operational role in this, and are primarily responsible for coordination with entities outside of their jurisdictions (e.g., the national disaster coordination center, foreign governments). With regard to the program's emphasis on improving information sharing systems, both countries have a common motive to support the program: to quickly exchange flood related information in a major flood. In the past, different municipalities and provinces were using different flood management systems and could not share information even within the same country. The initial project budget for Program VIKING was 1.6 million Euros. Because the program is partly financed by European funds, the products of the project (e.g., reports, software) are available for government agencies of other EU countries. So far, Ireland, Scotland, England, Poland, Czechia, Slovakia and Hungary have expressed interest in the products.

Program VIKING has been a successful project, according to multiple evaluation reports from both Dutch and German sides. Every stakeholder the research team has interviewed also considered it a successful project. For example, one of the project managers noted that they could obtain continued commitment from the participating agencies and additional funding to extend the collaboration efforts after the original project term (~2006) because their cross-border collaboration and DM ISD were successful. The program managers also believed that the collaboration between the Netherlands and Germany had significantly improved since Program VIKING started.

Another indicator of the success is the prestigious National Safety Award bestowed upon Program VIKING in 2006. In addition, several countries including Slovakia and Romania have decided to adopt FLIWAS [45]. Therefore, at least as of the time this study was conducted, Program VIKING was viewed as a good exemplar of successful cross-border DM and DM ISD project.

Analysis Results

To keep our research manageable, we demarcated the beginning of Program VIKING as the starting point of our analysis when the VIKING Alliance was first formed and took the leading position in the cross-border collaboration project. The alliance originally included only a small number of partners including the regional fire departments, the water boards, and some other organizations in the province of Gelderland, the Netherlands. On the German side, only one agency, Regional Government of Düsseldorf, participated in the initial alliance, resulting in an unbalanced alliance structure skewed toward the Netherlands. The Province of Gelderland was coordinating the Dutch partners, while the Region of Düsseldorf that recently took over the water management in Germany was representing various German municipalities and DMOs. The first project meetings clearly pointed out that a wider range of partnership that includes regional police departments, municipalities, and ambulance services is required in order to develop a more sustainable program and gain regional commitment for further collaboration. Some of the major contradictions in the initial Program VIKING structure existed between:

- The budget and work to be done (instrument vs. outcomes)
- Political/economic power and the wider range of stakeholders to be mobilized (instrument vs. community/rules)
- The potential partners to be included and the hierarchical structure of the organizations (stakeholders vs. community/division of duties)

The alliance responded to these challenges by actively lobbying high level officials (action) in the hierarchies of DMOs in both countries, in an effort to expand the alliance partnership (action-level goal).

Although the action-level goal of the lobbying actions may not look relevant to the system-level objective (i.e., facilitating cross-border DM) of Program VIKING, it turned out to be an important step that helped the alliance build its capacity to cope with the pressure in the cross-border collaboration environment (e.g., different cultures, institutional rules, & technological preferences) and effectively utilize distributed organizational knowledge [46] to achieve intended outcomes in the subsequent actions, which eventually led to successful system-level activity of the program.

Program VIKING can be viewed as a social structure created by a cross-border DM initiative, and thus the pre-existing environments in the Netherlands and Germany, as well as the purposeful strategic decision making of the initiative, determined the initial fate of how the program was to be governed, operated, and concluded. Nevertheless, the program, as an evolving activity system, justified its existence and extended its lifespan by serving the needs of the DM communities in both countries. Our analytical framework identified four key boundary objects: the governance structure of the program (i.e., VIKING Alliance), two information systems (i.e., FLIWAS and VIKING Cockpit), and cross-border DM exercise (e.g., HAGAR, HELGA, and ROAR). These were once artifacts of Program VIKING's activities, but became parts of the program itself and played pivotal roles in the successful development and continuation of the program. Using the AT terms, Table 4 outlines the significant activities from which the key boundary objects emerged and resolved contradictions in Program VIKING. More detailed descriptions of the key boundary objects are presented in the following subsections.

Table 4: Analysis results of key activities

Activity	Subject	Instruments	Environments (i.e., Community, Division of labor, Rule)	Outcome
Contradictions to be resolved				**(Later became B.O.)**

Transformation of cross-border governance structure (2001–2004)	Initial partners of VIKING Alliance	Project meetings (physical)	Dutch side: The Province of Gelderland and some regional agencies (e.g., fire depts., water boards, etc.)	New governance structure of VIKING Alliance (2004, February)
		Lobbying		
→Participation and resources to perform all expected collaboration activities			German side: The Region of Düsseldorf.	→Expanded & mandated participation
→Authority to mobilized all stakeholders			The Province of Gelderland coordinates other Dutch partners, while Region of Düsseldorf represents Water Mgmt. and (non-member) German agencies in the region.	→Symmetric distribution of authority and control across the border
			The roles of Police, municipalities, and ambulance services cannot be covered by the initial partners.	
			Agencies in one country must be controlled by a higher authority in the same country, but there was no clear hierarchy set for the Alliance.	
Development of IS for DM collaboration (2005, May-November)	Transformed VIKING Alliance (esp. steering committee)	Project meetings (physical) with the end-user groups and key stakeholders	An expanded set of alliance partners with incompatible IS.	FLIWAS (version 1 was first released in November 2005)
→Exceedingly fragmented ISs		Decentralized/ unorganized info sharing	All partners in one country are now under the leadership of a single program manager in that country.	-First possibility of cross-boundary information sharing

→Need to support agencies in two countries/ languages		EU Funding secured by the transparent governance structure		-Awareness of the potential of DMIS for effective flood management
→The deployment, access, and use of the IS must be easy for the large number of partners		ISD feasibility study		-VIKING Cockpit (first online release in June 2006)
Development of IS for ISD Collaboration	"	"	Increasing number of alliance partners (e.g., the Dutch National Ministry of Internal Affairs)	→Efficient cross-border info sharing, knowledge/ project management for collaborative ISD
(2006, January-June)		+	Prospect of extended funding by the European commission	Easy access to & evaluation of DM IS (FLIWAS online version)
→Inadequacy of physical project meetings for the large stakeholders		-Advanced knowledge/ project management technologies		
→Inadequacy of decentralized information sharing among the large stakeholders		Existing Internet infrastructure		
→Need easier access to FLIWAS		FLIWAS (web-invoked light version)		
Institutionalization of regular field exercises (November 2005-May 2009)	"	"	Increasing number of alliance partners & other stakeholders (spectators, potential partners/ adopters/investors)	Cross-border DM
→Need to evaluate the performance of cross-border DM collaboration, information sharing, and DMIS		+	Prospect of commercialization (consultation and software delivery) to other countries in Europe	Exercises (HAGAR exercises in 2005, HELGA exercises in 2006 ROAR exercises in 2008)
Need to demonstrate and promote the program for continued support		FLIWAS		→Demonstration of the successful progress & justification of the collaboration program

		VIKING Cockpit		→Empirical test and improvement of collaborative DM IS (FLIWAS) & DM performance
		Performance metrics		→Stronger sense of community/ partnership
		Independent consultants		
		Academic researchers		
		Exercise planning, advertisement, promotion		

Bharosa *et al.*

Bharosa *et al. Security Informatics* 2012 1:15, doi:10.1186/2190-8532-1-15

Governance Boundary Object: cross-border Governance Structure

One of the first actions taken by the VIKING alliance was the installment of governance structure in itself. (Figure 3) illustrates the governance structure of Program VIKING. According to the project meeting notes, two requirements were initially set for this governance structure. The first requirement was that the representatives of the water management agencies should have a formal position in the governance structure. According to one of the informants, the motive behind this was "without the representation of the end-user group in the governance structure the program would not attain the necessary level of commitment of all the necessary partners. Moreover, we had to incorporate the end-users in our group meetings in order to understand their information needs". The second requirement for the cross-border governance structure was that it should equally distribute the authority and control between Dutch and German partners. Consequently, the alliance appointed two program managers, one for the Dutch partners and one for the German partners, to mediate the partners at the national level. One of the program managers stated that "we needed to have two program managers because we wanted to gain full commitment of all the agencies involved. It would be very difficult for a Dutch program coordinator to organize and lead German program teams... the [dual-head] leadership was also expected to better appreciate the cultural differences and promote the trust in the program, in addition to securing commitments, from the both sides"

Figure 3: Governance structure of Program VIKING.

The creation of governance structure for the VIKING alliance can be interpreted as a goal-oriented purposeful action, from the AT perspective. VIKING alliance was the subject of the action that formalized their structure (outcome) by using various instruments like requirement analyses and project meetings. This action and the resulting outcome addressed a contradiction between the alliance partners (subject) and the (absence of) rules and specification of duties to coordinate them. The design of this governance structure was influenced by the existing structure (e.g., the organizational hierarchies among the alliance partners, power structure in the cross-border context) and yet could influence the following actions as an added structural factor of Program VIKING. According to the analysis framework, this is because the outcome became a boundary object for the stakeholders while it was transforming the structure to resolve the internal contradiction. In addition to developing a shared identity among the alliance partners-a necessary condition for a boundary object [41]-the dual-head leadership prevented potential conflicts emerging from cultural differences by separating control over the partners on each side of the border. The two program managers acted as spokespersons for partners in each country to ensure the internal consistency of the alliance within and across the border. The governance structure also assured commitments of key stakeholders by mandating their involvement in the governance structure. As a result, partners could understand their roles in cross-border DM operations and had a good overview of what was expected from them and from the other partners. In general, the governance structure promoted a high level of trust among the alliance partners by providing a sense of equality and transparency. Especially, the high level of transparency developed through the governance structure was critical for obtaining further EU funding for Program VIKING. Mandated participation of key stakeholders and dual-head leadership are two distinctive attributes of the self-evolved governance structure. In terms of authority, the appointed program managers were well known and experienced leaders in the DM community and provided with the legal authority to mobilize all alliance partners in each side of the border. As a result, the governance structure could successfully coordinate the roles, responsibilities and resources of the alliance partners without explicit rules and policies.

IS Boundary Object: information Systems for Collaborative Disaster Response

As one of the original focus areas of Program VIKING was cross-border information sharing, one deliverable expected from the program was an architecture that would interconnect existing ISs of various DMOs across the border. Nevertheless, development of a new IS was not planned in the project proposal. The steering committee conducted a study of the existing DMIS in both countries, in an effort to find the best way to integrate the ISs already in place. However, the analysis reports revealed that the existing IS not only lacked the functionality for flood management, but also were exceedingly fragmented with overlapping functionalities. Based on the analysis reports, the steering committee concluded that it would be more viable for the partners to develop an entirely new IS for cross-border DM, which was later named FLIWAS. The purpose of FLIWAS was to "ensure that all relief workers on both sides of the border have access to the same information systems and make decisions based on real-time data in a disaster situation." Thus, FLIWAS has strong a characteristic of decision support solutions with the capability to integrate data from heterogeneous sources and run simulations to predict possible flood scenarios. The FLIWAS design guidelines specify that: the application (1) must be multilingual (Dutch, German and English), (2) accessible on the web (invocation via a browser), (3) come in multiple versions (e.g., full vs. light, installable vs. online) with a modular-architecture, and (4) comply with open standards. Major functional modules developed to date include: flood visualization, flood level and risks prediction, evacuation scenario simulation, and emailing.

FLIWAS is designed to make use of existing measurement and flood forecast systems, flood warning plans, flood risk maps and disaster scenarios, as well as various geographical information systems (GIS) data [47]. Users can selectively install the modular components of FLIWAS to meet their specific needs, while their access to the different functions is controlled by a role based authentication system [47,48]. The system uses a client–server architecture, where an Internet browser or installed software agent can be used to access FLIWAS services running on a Linux-based server [45]. The GIS module is built on UMN-MapServer, and the databases can be built on PostgreSQL or Oracle [48,49]. (Figure 4) illustrates the FLIWAS-aided information flows between different Dutch and German DMOs.

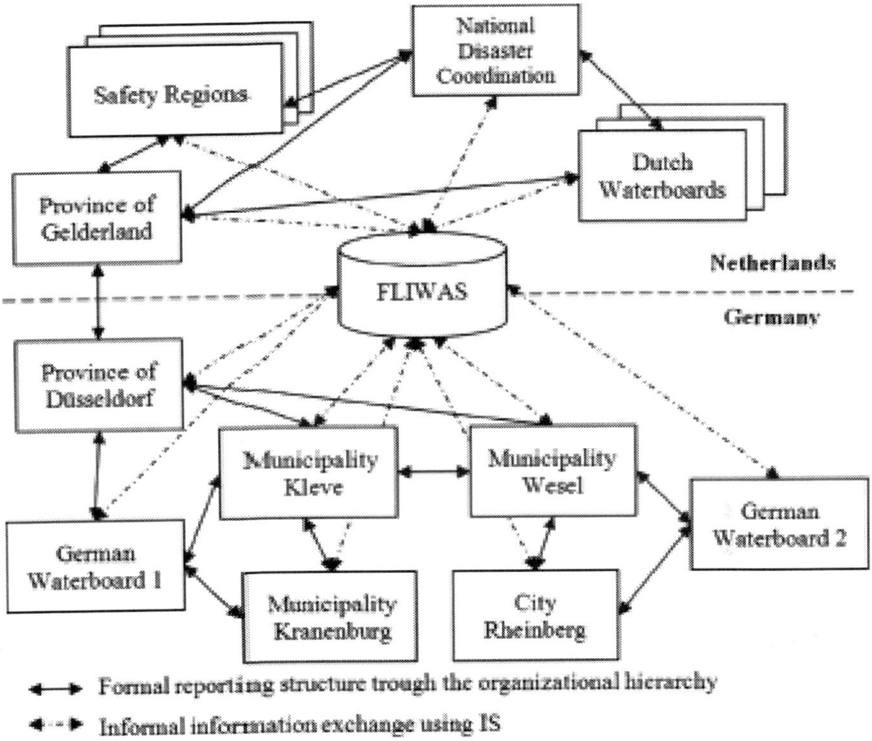

Figure 4: Information flows.

As shown in the figure, only one formal communication channel existed between province of Gelderland and Region of Düsseldorf at the international level, which had to mediate all cross-border information flows from/to the waterboards and regional DMOs. Even within a country, not every DMO had a direct formal connection with every other DMO. FLIWAS enabled DMOs to bypass the clogged formal communication channel and instantly exchange information with each other, achieving a shared situational awareness of the status of dikes, water level, and pumps during a flood.

IS Boundary Object: information Portal for Collaborative Project Management

Similar to the case of FLIWAS, the initial proposal for Program VIKING did not include any plan to develop an IS for the collaborative program itself. In the early stages of the program, the alliance partners shared information with each other via physical group meetings and presentations, printed documents, and emails. However, as the number of partners steadily increased, these means of information sharing quickly turned inefficient, and issues such as document versioning, format compatibility and security started to gain priority on the meeting agendum. After multiple requests from partners for a shared document repository, the steering committee decided to initiate the development of a collaboration portal, which resulted in a new IS named VIKING Cockpit. VIKING Cockpit collects and provides access to all essential information of Program VIKING. There were three initial requirements for the VIKING cockpit. First, this cockpit should be able to store a large amount of data in various formats, ranging from geographical maps to meeting reports. Second, access to this portal should be secured as the stored data may contain sensitive information. Finally, the portal should be available in both Dutch and German. Later, another requirement was added to the list: the portal should be able to invoke FLIWAS. Consequently, the VIKING Cockpit now has a link to a light version of FLIWAS and thus can play the role of a DMIS during disaster, in addition to the role of a collaboration portal for the program during a non-disaster time.

FLIWAS and the VIKING cockpit are two major, but originally unplanned IS artifacts of Program VIKING. Development of FLIWAS was initiated by an expansive learning-a realization that the original plan of action (i.e., interconnecting existing ISs) was not a good solution, given the available economic, organizational, and technical resources (instruments), for the system level activity (i.e., facilitate cross-border DM). Similarly, the VIKING Cockpit was developed to better fulfill the alliance partners' managerial needs for the collaborative project, rather than for cross-border DM operations. Since both ISs are expansive solutions designed to solve eminent contradictions in the existing structure (e.g., the large number of stakeholders and their legacy systems incapable of effective information sharing), many alliance partners quickly accepted these solutions. For example, while FLIWAS was still in the beta stage, two out of three Dutch Waterboards had already adopted it to exchange information with their neighboring German Waterboards. Such proactive early adoptions accelerated the IS-led transformation of the cross-border DM communication structure, as the practice instated an efficient direct information sharing network across the two countries, in addition to the existing hierarchical and formal reporting channel [46]. The link to FLIWAS in the VIKING Cockpit further boosted the responders' adoption of both ISs and secured their places in the institutional structure. Even though many DMOs have not officially adopted FLIWAS yet, they still can try FLIWAS via VIKING Cockpit anytime anywhere, which has made the learning and feedback structure of Program VIKING radically easy and simple. The enforced involvement of key stakeholders and use of open standard in the IS development process also helped lower resistance to the ISs and cultivated the emergence of a common identity among the alliance partners as they engaged in a common set of practices (i.e., development, evaluation, and revision of the ISs) [25]. Note that FLIWAS and VIKING Cockpit have gone through a cyclic process of evaluation and revision, through which the cross-border DM and DM ISD structures have also been transformed along the ISs.

Feedback Boundary Object: cross-Border Exercises

The other key boundary object that we have found is a series of large scale cross-border DM exercises. Three major exercises are: HAGAR (2005), HELGA (2006), and ROAR (2008). The explicit goals of the exercises were to improve (1) multi-agency collaboration between relief agencies on different echelons (strategic, tactical and operational levels) of response, (2) cross-border information sharing between the Dutch and German relief agencies, (3) information provisioning to the operational units during floods, and (4) developments of skills in using information systems for flood control. Apparently, the primary contradiction addressed by the exercises is the fact that cross-border DM structure, including the institutional rules, coordination plans, and operational procedures, cannot be examined under the normal condition due to the distributed DM responsibilities (division of duties). VIKING Alliance used its political and relational power in order to mobilize alliance partners and resources to conduct full-scale exercises, which provided the participants with chances to get acquainted with the cross-border DM structure, surrounding environments, and available ISs. ROAR had an additional objective to evaluate whether or not the newly developed DMIS (i.e., FLIWAS) were successful in meeting the requirements of the DMOs. One software developer stated that: "Even though we knew FLIWAS was not yet fully functional, we had to show our sponsors and users that FLIWAS was on a path to success, in order to build a foundation for survival of the program and prevent partners from exiting the alliance". Hence, ROAR was also used as a showcase and a test bed that allowed a large number of stakeholders to experience and evaluate the new DMIS.

The cross-border exercises made significant impacts on the collaborative DM activities in several ways. Since the exercises were recursive and large in scale, they became a widely recognized institutional mechanism for training and evaluation by mid-2007. Through the exercises, end-users of the newly developed FLIWAS could try the system in a simulated disaster environment, share their ideas with other end-users, and provide feedback to DMIS developers. In doing so, they could also improve their understanding of the roles and interdependencies of various DMOs in the larger context of cross-border DM operations. Thus, the institutionalized exercises became a knowledge creation and sharing mechanism, significantly improving the expansive learning capacity of Program VIKING. Cross-border exercises have also been great places to exhibit the progress of the project, as the exercise organizers invited not only the current alliance partners but also members of a wider community including media, academics, investors, and DM managers in other regions and countries. By demonstrating the progress of the exercise, the program could earn trust from its stakeholders and secure further involvements and commitments (e.g., partnership & budget) for the program activity. Table 5 summarizes the attributes and positive effects of the key boundary objects, which can answer not only "what" influenced the successful cross-border DM collaborative but also "how" and "why" those success factors were enacted in the particular social structure.

Table 5: Attributes and effects of key boundary objects

Key boundary object*	Attributes	Consequences (factors)
1. VIKING Alliance Governance Structure	·Expanded partnership	·Assured commitments of key stakeholders
	→Covers the full spectrum of necessary functions for collaborative DM & ISD.	·Improved understanding on the division of duties.
	·Mandated participation	·Prevented cultural conflicts
	→Institutionalized positions for key stakeholders (e.g., end-user groups) in the governance structure	·Ensured national-level consistency
	·Dual-head leadership	·Ensured political power to mobilize all alliance partners
	→Appoint an authoritative leader.	·Provided a sense of equality and transparency
	→Confer enough legal authority.	→Promote trust among partners
		→Secure further funding
2. Information Systems	·Convenient & secure access to all partners	·Created a new layer of direct communication links for DMOs.
A. FLIWAS	→Web-based	→Bypass hierarchical controls
B. VIKING Cockpit (Knowledge/Project Mgmt.)	·Modular multi-version	→Remove stove-piping
	→Flexible implementation options	→Remove bottleneck at the border
	·Multi-lingual	·Increase knowledge sharing/ creation for ISD
	·Neutral design	·Lowered adoption barrier
	→Use open standards	·Interested other EU states
	→Independent of national DMO structure.	·Increased acceptance of IS
	·Mandated involvement of all key stakeholders	·Increased interoperability
3. Feedback mechanism (Cross-border DM Exercises)	·Large scale to involve the full spectrum of	·Improved individuals' understanding of cross-border DM operations.
	→Partners	·Provided a built-in feedback mechanism
	·Recurrent exercise	→Offer DMIS evaluation opportunities
	·Invite potential stakeholders	→Encourage expansive learning
		·Demonstrated the progress of the ISD
		→Showcase for FLIWAS to EU states

		→Revamp internal interests, involvement, and commitment
		→Keep EU/external supports

* Conditions for a key boundary object in the analysis framework: 1) developed as an expansive solution, meaning it is adaptable and scalable to changing stakeholder needs, and 2) become an influential part of the activity system structure.

Bharosa *et al.*

Bharosa *et al.* *Security Informatics* 2012 1:15, doi:10.1 186/2190-8532-1-15

CONCLUSIONS

This research developed and applied an AT-based analytical framework to investigate key boundary objects in cross-border ISD for disaster management. We identified three types of boundaries objects that greatly contributed to the cross-border collaboration: 1) governance structure, 2) information systems and 3) recurring evaluation/feedback opportunities. The information systems category included two boundary objects: an easily accessible DMIS prototype and an online portal for project management. The selective institutionalizations of these key boundary objects helped the participants to overcome various contradictions existed in the socio-technical system of collaborative cross-border disaster management. Specifically, the uncertainty about the collaboration project's future (i.e., lifespan), the limited information sharing capability in the DM community, and the wide geographic and functional dispersions of the collaboration partners were alleviated by the key boundary objects.

Provided the circumstances are similar, managers of cross-border DM projects are recommended to take a strategy that develops and institutionalizes the above mentioned key boundary objects. It is not the creation of boundary objects however, but the attributes possessed by those boundary objects that actually determine the fate of a project. For example, each participating country must have an authoritative and trusted representative in a trans-national governance structure, if the governance structure were to play a key role in the cross-border

DM project. For an information system to be a key boundary object, it should allow all potential users convenient and secure access, flexible implementation options, and power to overcome the limitation in the existing DMO structure (e.g., formal reporting channels in the national incident management system). DM exercises can also be a key boundary object if they involve the full spectrum of stakeholders and offer evaluation and feedback opportunities in regular bases. The managers of cross-border DM projects are also expected to constantly monitor and analyze, using our analytical framework, the socio-technical structure of their project. They should look for structural contradictions and devise a solution that can be implemented with existing means and resources. If such a solution has the characteristics of boundary objects, the managers should promote the solution to be a more permanent part of the project structure even after the target contradictions are eliminated. Adding a clear and useful boundary objects will further improve cross-organizational interactions, commitments and appreciation of the project activities and outcomes.

This study also contributes to the academic community by developing a framework that analyzes the process through which material and social objects become interwoven in a cross-border DM collaboration system. This AT-based framework can help researchers identify important key boundary objects that can overcome the structural limitations and challenging conditions of large-scale DM collaboration projects. This research is, to our best knowledge, the first to employ AT for an investigation of cross-border DM ISD. The principles of AT (e.g., mediation by objects, historicity and expansion) make this theory suitable for studying a cross-border DM ISD context. By accounting for the time component (i.e., historicity) in AT, our analytical framework recognizes the possible feedback loop between pre-existing social structure and goal-oriented purposeful actions. Accordingly, the proposed analytical framework can suggest potential causal links between the success factors for cross-border DM collaboration and the attributes of key boundary objects

While this study examines a cross-border ISD in the DM domain, the findings can provide useful insights to researchers in other domains as well. Many private-sector companies seek to establish collaborative relationships for new opportunities or threats. When such a collaborative relationship spans across multiple cultures, legal jurisdictions, and an extended period of time, the key boundary objects identified in this paper can be developed to induce a similar positive influence in a new context. Our framework will be particularly useful when the system being analyzed is a complex and persistent social structure that involves a large number of stakeholders. This study is based on a single ethnographic study and presents only a partial view of cross-border DM ISD collaboration. Since we studied two countries in Western Europe, some cultural similarities might have affected our findings. Future research may examine relationships between cultural distance and collaboration success in cross-border DM ISD. Although our qualitative approach provides a deep understanding of the processes and issues in question, the findings require quantitative validation. Therefore, developing and validating a quantitatively testable model of cross-border DM ISD success will be the next step of this research stream. Of course, more qualitative studies in the cross-border DM ISD collaboration area will enrich the small pool of testable hypotheses.

AUTHORS' CONTRIBUTIONS

All four authors equally contributed to this study. They were deeply involved in every step of the study including research design, literature review, data collection & analysis, paper write-up, and multiple cycles of internal revisions phases. All authors read and approved the final manuscript.

ACKNOWLEDGEMENTS

The research of the 2nd and 4th authors has been supported by US National Science Foundation under grant # IIS-0809186. The research of the fourth author has also been funded in part by Sogang Business School's World Class University Project (R31-20002), funded by Korea Research Foundation as well as by the Sogang University Research Grant of 2011. The usual disclaimer applies. The authors appreciate the Dutch disaster management personnel who provided their valuable opinions and the ROAR organizers, Program VIKING, for their cooperation and support for our research.

REFERENCES

1. Akmanligil M, Palvia P: Strategies for global information systems development. *Inf. Manage.* 2004, 42:45-59.
2. Urs SR, Raghavan KS: Vidyanidhi: Indian digital library of electronic theses. *Commun. ACM* 2001, 44:88-89.
3. Bharosa N, Lee J, Janssen M: Challenges and obstacles in information sharing and coordination during multi-agency disaster response: propositions from field exercises. *Inf. Syst. Front.* 2010, 12:49-65.
4. Turoff M, Chumer M, Van De Walle B, Yao X: The design of a dynamic emergency response management information system (DERMIS). *J Inf Technology Theory and Application* 2004, 5:1-35.
5. Fedorowicz J, Gogan JL: Reinvention of interorganizational systems. *Inf. Syst. Front.* 2010, 12:81-95.
6. Sagun A, Bouchlaghem D, Anumba CJ: A scenario-based study on information flow and collaboration patterns in disaster management. *Disasters* 2009, 33:214-238.
7. Fedorowicz J, Gogan JL, Williams CB: A collaborative network for first responders: lessons from the CapWIN case. *Gov. Inf. Q.* 2007, 24:785-807.
8. Xu JJ, Chen H: Fighting organized crimes: using shortest-path algorithms to identify associations in criminal networks. *Decis. Support. Syst.* 2004, 38:473-487.
9. Romano N Jr, Pick J, Roztocki N: A motivational model for technology-supported cross-organizational and cross-border collaboration. *Eur. J. Inf. Syst.* 2010, 19:117-133.
10. Careem M, Silva C, Silva R, Raschid L, Weerawarana S: *Sahana: Overview of a disaster management system*. In IEEE international conference on information and automation, Colombo, Sri Lanka; 2006:361.
11. Engestrom Y: Activity theory and individual and social transformation. In*Perspectives on activity theory*. Edited by Engeström Y, Miettinen R, Punamäki RL. University Press, Cambridge; 1999:19-38.
12. Engeström Y: *Learning by expanding: An activity-theoretical approach to developmental research*. Orienta-Konsultit, Helsinki;

1987.

13. Nardi B: *Context and consciousness: Activity theory and human computer interaction*. MIT Press, Cambridge; 1996.
14. Vygotsky LS: *Mind in Society: The development of higher psychological processes*. Harvard University Press, Cambridge; 1978.
15. Engeström Y, Miettinen R, Punamäki RL: *Perspectives on activity theory (learning in doing: social, cognitive and computational perspectives)*. Edited by Punamäki RL. University Press, Cambridge; 1999.
16. Bertelsen OW, Bodker S: Activity theory. In *HCI models theories, and frameworks: Toward a multidisciplinary science*. Edited by Caroll JM. Morgan Kaufmann; 2003:291-324.
17. Kuutti K: Activity theory as a potential framework for human computer interaction research. In *Context and consciousness: Activity theory and human-computer interaction*. Edited by Nardi BA. Massachussetts Institute of Technology, Cambridge; 1995:17-44.
18. Chen R, Sharman R, Chakravarti N, Rao HR, Upadhyaya SJ: Emergency response information system interoperability: development of chemical incident response data model. *J. Assoc. Inf. Syst.* 2008, 9:200-230.
19. Chen R, Sharman R, Rao HR, Upadhyaya S: An exploration of coordination in emergency response management. *Commun. ACM* 2008, 51:66-73.
20. Shankar D, Agrawal M, Rao HR: Emergency response to mumbai terror attacks: An activity theory analysis. In *Cyber security, cyber crime and cyber forensics: Applications and perspectives*. Edited by Santanam R, Sethumadhavan M, Virendra M. IGI Global, Hershey, PA; 2011:46-58.
21. Yates J, Orlikowski WJ: Genres of organizational communication: a structurational approach to studying communication and media. *Acad. Manage. Rev.* 1992, 17:299-326.
22. Orlikowski WJ: Using technology and constituting structures: a practice lens for studying technology in organizations. *Organ.*

Sci. 2000, 11:404-428.

23. Orlikowski WJ: The duality of technology: rethinking the concept of technology in organizations. *Organ. Sci.* 1992, 3:398-427.
24. Tushman M, Romanelli E, Staw BM: Organizational evolution: A metamorphosis model of convergence and reorientation. In *Research in organizational behavior*. Edited by Cummings LL, Staw BM. JAI Press, Greenwich; 1985:171-222.
25. Levina N, Vaas E: The emergence of boundary spanning competence in practice: implications for implementation and use of information systems. *MIS Q.* 2005, 29:335-363.
26. Engestrom Y: Activity theory as a framework for analyzing and redesigning work. *Ergonomics* 2000, 43:960-974.
27. Gal U, Lyytinen K, Yoo Y: The dynamics of IT boundary objects, information infrastructures, and organisational identities: the introduction of 3D modelling technologies into the architecture, engineering, and construction industry. *Eur. J. Inf. Syst.* 2008, 17:290-304.
28. Star SL, Griesemer JR: Institutional ecology, 'translations' and boundary objects: amateurs and professionals in Berkeley's museum of vertebrate zoology, 1907–1939. *Soc. Stud. Sci.* 1989, 19:387-420.
29. Hussenot A, Missonier S: A deeper understanding of evolution of the role of the object in organizational process: The concept of "mediation object". *J. Organ. Chang. Manag.* 2010, 23:269-286.
30. Carlile PR: A pragmatic view of knowledge and boundaries: boundary objects in new product development. *Organ. Sci.* 2002, 13:442-455.
31. Lindgren R, Andersson M, Henfridsson O: Multi-contextuality in boundary-spanning practices. *Inf. Syst. J.* 2008, 18:641-661.
32. Levina N: Collaborating on multiparty information systems development projects: a collective reflection-in-action view. *Inf. Syst. Res.* 2005, 16:109-130.
33. Strauss A, Corbin J: *Basics of qualitative research*. Sage, Newbury Park; 1990.
34. Benbasat I, Goldstein D, Mead M: The case research strategy in studies of information systems. *MIS Q.* 1987, 11:369-386.

35. Pettigrew AM: Longitudinal field research on change: theory and practice. *Organ. Sci.* 1990, 1:267-271.
36. Merton RK, Fiske M, Kendall PL: *The focused interview: A manual of problems and procedures*. 2nd edition. Free Press, New York; 1990.
37. Spradley JP: *The ethnographic interview*. Holt, Rinehart & Winston, New York; 1979.
38. McCall CJ, Simmons JL: *Issues in participant observation: A text and reader*. Addison-Wesley Publishing Company, Reading, MA; 1969.
39. Kluchkohn FR: The participant observer technique in small communities. *The America J Sociology* 1940, 46:331-343.
40. Yakura EK: Charting time: timelines as temporal boundary objects. *Acad. Manage. J.* 2002, 45:956-970.
41. Levina N, Vaast E: The emergence of boundary spanning competence in practice: implications for implementation and use of information systems. *MIS Q.* 2005, 29:335-363.
42. Chen R, Coles J, Lee J, Rao HR: Emergency communication and system design: The case of Indian ocean tsunami. In *International conference on information and communication technologies and development (ICTD); Apr 17–19, 2009*. Doha, Qatar; 2009:300-309.
43. Lacity MC, Janson MA: understanding qualitative data: a framework of text analysis methods. *J. Manag. Inf. Syst.* 1994, 11:137-155.
44. Hofstede G: *Culture's consequences, comparing values, behaviors, institutions, and organizations across nations*. Sage Publications, Thousand Oaks CA; 2001.
45. Vinke de Kruijf J: The role of Dutch expertise in romanian water projects-case study 'pilot implementation FLIWAS in Banat region, Romania'. In *Book the role of Dutch expertise in romanian water projects -case study 'pilot implementation FLIWAS in Banat region, Romania'*. Department of Water Engineering and Management/Twente Centre for Studies in Technology and Sustainable Development; 2011:91.
46. Mahesh K, Suresh JK: Knowledge criteria for organization design. *J. Knowl. Manag.* 2009, 13:41-51.

47. De Gooijer K: FLIWAS, the right information at the right place at the right time for the right persons to take the right decision. *BALWOIS 2010; May 25–29 Ohrid, Republic of Macedonia* 2010.

48. Langkamp EJ, Wentholt LR, Pengel BE, De Gooijer C, Flikweert J, Langkamp EJ, Wentholt LR, Pengel BE, De Gooijer C, Flikweert JJ: *Floods, from defence to management.* Edited by Van A. Taylor & Francis Group, London; 2005:305-308.

49. Gretzschel M, Jüpner R, Grafe M, Leiner R: *Application of flood management systems in germany.* The First European IAHR Congress, Edinburgh, UK; 2010.

Chapter 7

Global Database on Large Magnitude Explosive Volcanic Eruptions (LaMEVE)

Helen Sian Crosweller[1], Baneet Arora[2], Sarah Krystyna Brown[1], Elizabeth Cottrell[3], Natalia Irma Deligne[1,4], Natalie Ortiz Guerrero[1,5,6], Laura Hobbs[1,7], Koji Kiyosugi[8], Susan Clare Loughlin+, Jonathan Lowndes[9], Martin Nayembil[2], Lee Siebert[3], Robert Stephen John Sparks[1], Shinji Takarada[10], and Edward Venzke[3]

[1]School of Earth Sciences, University of Bristol, Queens Road, Bristol, BS8 1RJ, UK

[2]British Geological Survey, Nicker Hill, Keyworth, Nottingham,, NG12 5GG, UK

[3]Smithsonian Institution, National Museum of Natural History, Washington, DC, 20560, USA

[4]Current address: GNS Science, Avalon, Lower Hutt, 5040, New Zealand

[5]Current address: Faculty of Engineering, Universidad Mariana, San Juan de Pasto, Nariño, Colombia

[6]Current address: Instituto de GeofÃsica, Universidad Nacional Autónoma de México, Ciudad Universitaria, C.P. 04510, Coyoacán, Distrito Federal, México

[7]Current address: Lancaster Environment Centre, Lancaster University, Lancaster,, LA1 4YQ, UK

[8]Department of Geology, University of South Florida, 4202 East Fowler Avenue, Tampa, Florida, 33620, USA

[9]British Geological Survey, Murchison House, West Mains Road, Edinburgh,, EH9 3LA, UK

[10]Geological Survey of Japan, AIST, Site 7, 1-1-1, Higashi, Tsukuba, Ibaraki, 305-8567, Japan

ABSTRACT

To facilitate the assessment of hazards and risk from volcanoes, we have created a comprehensive global database of Quaternary Large Magnitude Explosive Volcanic Eruptions (LaMEVE). This forms part of the larger Volcanic Global Risk Identification and Analysis Project (VOGRIPA), and also forms part of the Global Volcano Model (GVM) initiative (http://www.globalvolcanomodel.org *webcite*). A flexible search tool allows users to select data on a global, regional or local scale; the selected data can be downloaded into a spreadsheet. The database is publically available online at http://www.bgs.ac.uk/vogripa *webcite* and currently contains information on nearly 3,000 volcanoes and over 1,800 Quaternary eruption records. Not all volcanoes currently have eruptions associated with them but have been included to allow for easy expansion of the database as more data are found. Data fields include: magnitude, Volcanic Explosivity Index (VEI), deposit volumes, eruption dates, and rock type. The scientific community is invited to contribute new data and also alert the database manager to potentially incorrect data. Whilst the database currently focuses only on large magnitude eruptions, it will be expanded to include data specifically relating to the principal volcanic hazards (e.g. pyroclastic flows, tephra fall, lahars, debris avalanches, ballistics), as well as vulnerability (e.g. population figures, building type) to facilitate risk assessments of future eruptions.

BACKGROUND

Explosive volcanism represents a key Earth process that transfers heat and mass from the Earth's interior to the surface. Its products contribute constituents to global geochemical cycles, affecting the near surface geosphere, hydrosphere and atmosphere. Volcanic eruptions have the potential to cause loss of life, disrupt air traffic, impact the climate and significantly alter the surrounding landscape. Knowledge of magnitude-frequency relationships is poorly constrained at a majority of volcanoes but is key for assessing environmental and societal impacts of volcanism on global, regional and local scales. We have created a database of Large Magnitude Explosive Volcanic Eruptions (LaMEVE) to support these assessments and provide basic information on global explosive volcanism.

The LaMEVE database contains the nearly 3,000 Quaternary volcanoes catalogued by the Smithsonian's Global Volcanism Program and over 1,800 explosive eruption records spanning the last 1.8 My. Not all volcanoes currently have eruptions associated with them but have been included to allow for easy expansion of the database as more data are found. The ultimate objective is to go back to the start of the Quaternary at its current definition of 2.58 Ma, but at present the database is limited to the previous definition of 1.8 Ma.

The database, accessible online at http://www.bgs.ac.uk/vogripa *webcite*, primarily consists of data on eruption magnitude, age and source volcano location. The data were collected from published literature; principally Volcanoes of the World by Siebert et al. (2010) and journal articles, but also other online databases, books, public reports, conference proceedings, etc. This database should be of particular interest to volcanology researchers, hazard modellers and civil authorities responsible for crisis management.

There are 3 criteria an eruption must satisfy in order to be included in the database:

- It must be magnitude (M) (Pyle, 2000) or VEI (Volcanic Explosivity Index) (Newhall and Self,1982) of 4 or above.
- It must be dated.
- It must be from a known source volcano.

The database also contains information on deposit type, deposit volumes, eruption intensity, rock type classification, data errors and uncertainties with indices of data reliability. The database is currently restricted to larger magnitude eruptions because they have large hazard footprints that can threaten large populations. They are difficult to predict and occur less frequently than smaller eruptions. Some of the largest historic eruptions occur at volcanoes presumed dormant; as such, preparation and response are often poor. These larger eruptions are typically responsible for the greatest loss of life in the historical period (Siebert et al., 2010, pp. 41).

There exist two other databases similar in scope to LaMEVE: that of the Smithsonian Institution's Global Volcanism Program (GVP, http://www.volcano.si.edu *webcite*; Siebert et al. 2010) and the database by Mason et al. (2004) with eruptions of M≥8. LaMEVE, however, is distinctly different from these two. The database maintained by the Smithsonian's Global Volcanism Program is the leading source of global information on volcanic eruptions at all magnitude scales; however, it currently focuses on Holocene eruptions (the last 10,000 years). Analysis of large magnitude eruptions over this timescale has shown that this is an insufficient amount of time for adequate sampling of the largest eruptions (M>6.5) (Deligne et al., 2010). The database of Mason et al. (2004) concentrated on the very largest magnitude eruptions, which are considered to have the greatest potential for global repercussions. LaMEVE covers all eruptions of M≥4, as these eruptions still have the capacity to cause considerable regional and local disruption.

LaMEVE is the first component of the VOGRIPA (Volcanic Global Risk Identification and Analysis Project) database of volcanic hazards, which is being developed as part of the Global Volcano Model (GVM, http://www.globalvolcanomodel.org). The long-term aim of GVM is to have a global source of freely available information on volcanic hazards and volcanism produced to agreed standards and protocols.

This article discusses the structure and fields of the LaMEVE database, including its links with other databases, some of the difficulties encountered whilst constructing the database, how these issues were dealt with, perceived uses, and planned future developments. The paper also outlines planned steps to make the database accessible as well as further development ideas, including the ability for the scientific community to add new data.

DATABASE DESIGN

The LaMEVE database uses Oracle 10g R2 as a platform and contains 2 main tables and 8 secondary tables with additional look-up tables for codes used. The data format closely follows that of the Smithsonian's GVP database with many of the same data fields and codes, although the underlying structure of LaMEVE is not identical. The Global Volcanism Program's database is currently undergoing restructuring (Cottrell et al., in prep.) to allow public access and to better coordinate with the needs of VOGRIPA and other international databases. The LaMEVE database structure is outlined in Figure 1 with a full schema included in a supplementary file.

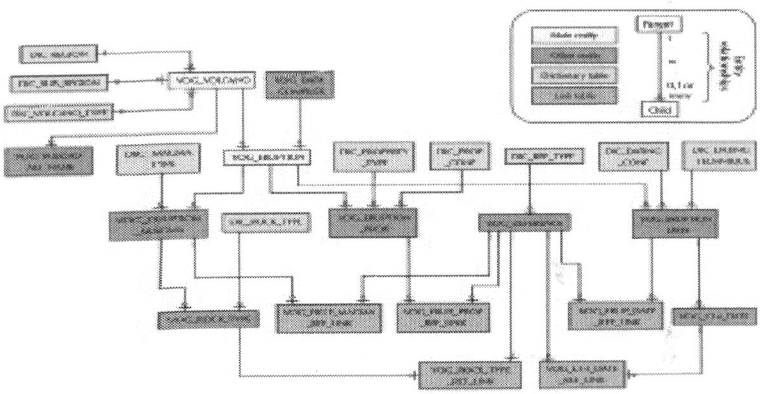

Figure 1: Conceptual diagram showing the current structure of the LaMEVE database. All the tables are linked via a one-to-many relationship whereby a single record in the parent table can relate to zero, one or many records in the child table. In instances where there is a many-to-many relationship between entities, this is resolved by using a linking table with a one-to-many relationship to each of the two entities. Dictionaries act as look-up tables for codes used in the main and other entities.

The following section is split into sub-sections based on the tables within the database. The two main tables are 'Volcano' and 'Eruption,' with the other tables linking to these; therefore the following discussion will be separated into these two main headings. Volcano attributes include spatial location and volcano type, while eruption attributes include timeframe data and quantitative eruption size characteristics. This database was structured using the 'normalisation' method, which is a process of removing duplication in the data, therefore minimising redundancy and the chance of entering inconsistent data (Connolly and Begg, 2005; Powell 2006).

Volcano

The Volcano table contains information on almost 3,000 volcanoes from around the world. The information is derived from the Smithsonian Institution's GVP database and includes volcanoes that are known or suspected to have erupted within the Holocene or late Pleistocene. Also included are some volcanoes now in the fumarolic stage, which signifies either a long pause between eruptions or residual degassing of a volcano that has not erupted for many thousands of years (Siebert et al., 2010, p3).

Data fields include volcano name and Smithsonian volcano number (VNUM), which are sanctioned by the International Association of Volcanology and Chemistry of the Earth's Interior (IAVCEI). If data are found on an eruption from a volcano not currently in the GVP database, the Smithsonian will be responsible for generating a new VNUM, which we would then use to add the data to our database. Additional fields include a unique table identifier[a], latitude and longitude coordinates, region and sub-region codes and volcano type as assigned by the Smithsonian.

[a]An automatically generated unique code (typically numerical) assigned to each row in the table to enable cross-referencing to each record.

Volcano Alternative Name

Alternative names, also sourced from the Smithsonian's database, are listed. These can be synonyms (i.e. different names for the same volcano), some of which are different spellings of the same name (e.g. Vulsini /Volsini) while others are significantly different from the official name (e.g. Santorini / Thera). They can also include names of features (e.g. cones or calderas). Although eruption records are linked via the main volcano name, the alternative names can be used to retrieve data from the appropriate volcano records.

Eruption

The Eruption table lists the unique identifier ID field (required in every table) and the Unit Name, of which there may be more than one listed. This records how the eruptive deposit is referred to in the literature, and provides a way of ensuring that the data all relate to the same eruption. The other columns in the Eruption table are actually links to other data tables (called 'foreign keys'), discussed below. Foreign keys are joined to the main eruption table via a one-to-many relationship allowing multiple pieces of data relating to one eruption to be reported, e.g. more than one date for the eruption from different sources. There is no judgement on what data are included in the database; the user can therefore use their own discretion regarding which data they wish to use for their own analyses.

Eruption Properties

This is a data-rich table which reports the following eruption characteristics, described separately below:

- VEI
- Magnitude
- Intensity
- Column height
- Tephra fall deposit volume and DRE (Dense Rock Equivalent) volume
- Ignimbrite[b] volume and DRE volume
- Primary intra-caldera material deposit volume and DRE volume
- Bulk deposit volume and DRE volume

[b]'Ignimbrite' refers to both pyroclastic flow and pyroclastic surge deposits.

Whilst this is a large amount of data to store in one table, these are all interrelated and require the same fields, namely a value and, where appropriate, either an error or qualifier (e.g. less than, greater than) or both. These quantitative characteristics are grouped into 3 categories to facilitate data handling: eruption size, deposit volume and DRE volume. The source of each data point is labelled as either 'literature', in which case it is linked to the reference source, 'calculated' or 'assumed' (see section 2.2.1.5 for more information).

VEI

The Volcanic Explosivity Index (VEI) devised by Newhall and Self (1982) describes the size of explosive eruptions; values range from 0 (gentle) to 8 (colossal). VEI is reported in the Smithsonian Institution's GVP database and is the most widely reported eruption descriptor in volcanology. These data are used wherever available so as to avoid disparity between the databases. As there is no single piece of information that can fully describe the character of an explosive eruption, the VEI incorporates multiple criteria for assigning a value. The VEI is a semi-quantitative scale based on 8 criteria, but in most cases is based principally on the erupted tephra volume (ancient eruptions) or a combination of plume height and eruptive volume (observed eruptions). This enables the VEI to be assigned from both quantitative and qualitative data, even if some information is missing, allowing a great number of eruptions to be classified. However, the VEI implicitly assumes that eruption magnitude and intensity are related, which is not necessarily true (Carey and Sigurdsson, 1989; Pyle, 2000). The LaMEVE database records VEI, magnitude (M) and intensity (I), so the relationships between these parameters can be assessed; this topic is discussed in this paper's companion.

Magnitude

Magnitude is the preferred measure of eruption size used in the LaMEVE database. If not reported in the literature, it is calculated using the formula of Pyle (2000) with reference to deposit volumes:

$$M = log_{10}[erupted\ mass(kg)] - 7 \quad (1)$$

which is equivalent to:

$$M = log_{10}\left[DRE(m^3) * magma\ density(kg/m^3)\right] - 7 \quad (2)$$

This measure has the added advantage over the VEI of being a more precise measure of eruption size. Typical estimates of volumes justify calculating magnitudes to the first decimal place in the database, allowing more discrimination. It is a quantitative parameter rather than a qualitative indicator that can be used to estimate magnitude-frequency relationships. The largest explosive eruptions on Earth are ~M=9 and so comparable to the magnitude of the largest earthquakes.

For eruptions with an assigned VEI value but no volume data or magnitude reported in the literature, eruption magnitude is assigned according to the relationship found between these two variables. (Figure 2 shows that, for 90% of eruptions in the VEI 4–6 range, the associated magnitude is within the same range (i.e. for VEI 5, 90% of the magnitudes are between 5 and 6). There is more variability for VEI 7 and 8 but the median value is still within the same magnitude range. Since most VEI values are determined from volumes of eruption tephra the correlation is as expected but reassuring. Table 1 lists the mean magnitudes associated with each VEI value, which are then entered into the database if a magnitude has not been reported in the literature, and are labelled as 'calculated.'

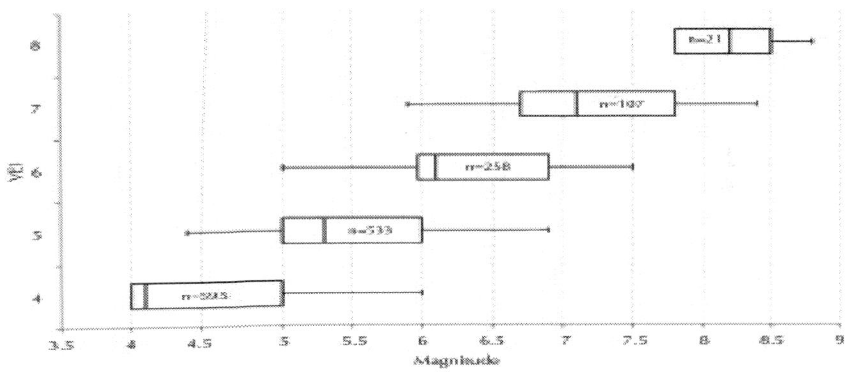

Figure 2: Box and whisker plot of VEI versus Magnitude. Boxes represent the range of the 5th and 95th percentiles with the dividing line being the median. Whiskers represent the minimum and maximum values. N values inside the boxes are the number of eruption records at that VEI.

Table 1: The minimum and maximum magnitude values for each VEI classification in the LaMEVE database as well as the mean magnitude with associated variance

VEI	Count	Min. mag.	Max. mag.	Mean mag.	Variance
4	595	4	6	4.4	0.14
5	533	4.4	6.9	5.3	0.13
6	258	5	7.5	6.3	0.13
7	107	5.9	8.4	7.2	0.14
8	21	7.8	8.8	8.2	0.06

The mean magnitude value is entered as an estimate of magnitude when only VEI is reported.

Crosweller et al.

Crosweller et al. Journal of Applied Volcanology 2012 1:4, doi:10.1186/2191-5040-1-4

Since VEI is, in the majority of cases, based on volume then Magnitudes in the range 4.0 to 4.9 should be equivalent to VEI 4. Table 2 shows that >80% of eruption records in the LaMEVE database have a magnitude value consistent with the VEI classification. The percentage of misclassifications (i.e. where Magnitude and VEI are not consistent with each other) is modest. Therefore, where VEI is missing but a magnitude value is reported, the rounded down magnitude value is entered for the VEI and labelled as 'calculated' (e.g. M=4.7 becomes VEI=4). Both values are included in the database to allow original source values to be reported wherever possible.

Table 2: Number and percentage of records in the LaMEVE database where magnitude values are either less than, the same as, or greater than the VEI classification

VEI	M <VEI		M =VEI		M ≥VEI+1	
4	0	0%	553	92.9%	42	7.1%
5	18	3.4%	483	90.6%	32	6%
6	14	5.4%	237	91.9%	7	2.7%
7	14	13.1%	89	83.2%	4	3.7%
8	2	9.5%	19	90.5%	0	0%

Crosweller et al.

Crosweller et al. Journal of Applied Volcanology 2012 1:4, doi:10.1186/2191-5040-1-4

Intensity and Column Height

Intensity is the rate at which magma is discharged and is primarily dependent on the pressure gradient between the magma chamber and the surface, magma viscosity, volatile content, and conduit dimensions (Carey and Sigurdsson, 1989). Intensity is calculated using the mass eruption rate as defined by Pyle (2000):

$$I = log_{10}[mass\ eruption\ rate(kg/s)] + 3 \tag{3}$$

It has been established that, for explosive eruptions, column height is related to intensity (Wilson et al., 1978; Sparks et al., 1997), reflecting the rate of thermal energy transfer into the atmosphere. Estimated column heights (in km) are determined either from direct observations for some historical eruptions, or by using the maximum clast size dispersal method of Carey and Sparks (1986). We note that this method has been recently updated by Burden et al. (2011), but has not yet been applied to the data in this database.

Deposit Volume and DRE

Where available, three distinct deposit volumes are recorded (tephra fall, ignimbrite, and primary intra-caldera material), which combined constitutes the bulk deposit volume. Volumes of associated lava flows are omitted from the database and excluded from calculations, thus given magnitudes are for the explosive phases of an eruption only and do not necessarily represent the magnitude of the entire eruption. The Dense Rock Equivalent (DRE) volumes are also stored. DRE corresponds to the unvesiculated erupted magma volume, i.e. the pre-eruption magma's volume. Where it is not reported in the literature, DRE is calculated using the following equation:

$$DRE(km^3) = \frac{tephra\ vol(km^3) * tephra\ density(kg/m^3)}{magma\ density(kg/m^3)} \tag{4}$$

Tephra density, unless otherwise reported, is assumed to be 1000 kg/m^3. Magma density varies according to the magma type (see below). For some records estimates and measurements of tephra deposit density are given; in these cases the deposit-specific literature value is used.

Data Quality Indicator

There is notable variation in the level of detail and consistency with which data are reported in the literature as well as in the clarity of explanation regarding data collection. This can be problematic when entering data into a database which follows a standardised format resulting in discrepancies between the reliability of different records which can be subtle. Therefore a 'data quality' indicator (ranging from 0 to 3) has been assigned to each eruption date and volume according to the criteria listed in Tables 3 and 4. This provides the user with a simple initial assessment of data reliability, and a way to exclude less reliable data from analyses. A full assessment of reliability should involve user evaluation of the original source information.

Table 3: Criteria for applying data quality indicators to eruption dates

DATING DATA QUALITY INDICATOR	
VALUE	**DESCRIPTION**
0	Only qualitative data provided (e.g. Holocene).
1	Date reported in the literature without error and or dating method not specified OR stratigraphically dated OR reported errors >30% OR unit has multiple published dates that differ by >30%.
2	Single date (historical or geochronology) with a method and errors cited OR unit has multiple published dates that differ by 10-30% OR historical ages uncertain by more than one year.
3	Single precise historical date within a year OR multiple published dates that differ by <10%.

Assigning data quality indicators to eruption dates permits users to prioritise superior analytical techniques. For multiple ages the above criteria are thus applied to those ages where there is no basis for assigning one age or method as superior to another.

Crosweller et al.

Crosweller et al. Journal of Applied Volcanology 2012 1:4, doi:10.1186/2191-5040-1-4

Table 4: Criteria for applying data quality indicators to volume data

VOLUME DATA QUALITY INDICATOR	
VALUE	**DESCRIPTION**
0	No isopach map or evidence to support cited volume OR assumed value.
1	Simple isopach map or thickness data (< 10 measurements) OR volume calculation with no methodology given OR volume derived from method other than isopach map (e.g. size of caldera; VEI estimate; duration and intensity of column heights).
2	Isopach map drawn from >10 and <30 measured thicknesses with calculation method described.
3	Isopach map drawn from >30 measured thicknesses with calculation method described.

There is no standardised method for constructing isopach maps from thickness data and several methods of extrapolation beyond the areas with data have been devised. The diversity of methods means there can be significant volume differences even with the same datasets when different methods are applied. Volumes are reported directly from sources using whichever method they applied, but preference is given to those calculated using the Pyle or Fierstein methods (Pyle 2000, Fierstein and Nathenson 1992). However, there are other methods and pre-1989 literature with various ways of integrating the volume from isopach maps. Some volumes include pyroclastic flow deposits where isopach maps are not necessarily readily drawn. Here the number of thickness measurements as above is used to assign a volume data quality indicator.

Crosweller et al.

Crosweller et al. Journal of Applied Volcanology 2012 1:4, doi:10.1186/2191-5040-1-4

We assign the lowest data quality level to those magnitudes and volumes that are 'assumed.' These are cases where quantitative data are inferred from qualitative descriptions. For example, an eruption might be described as 'Plinian' or 'caldera forming'; such an eruption is clearly explosive and therefore eligible for inclusion in the database (e.g. 'Tegalsruni' eruption unit from Merapi). The Smithsonian classifies these events as VEI 4 or larger, but does not assign a specific VEI. A magnitude of ≥4 is thus assigned to these eruptions in the LaMEVE database with the derivation method listed as 'assumed'. This permits the event to be included in a count of explosive eruptions despite an unknown actual eruption size. Eruption size and corresponding data quality can be updated as new information becomes available. When an assumed VEI/magnitude is entered the following corresponding assumed volumes are entered, as from Newhall and Self (1982):

- VEI 4: tephra volume ≥ 0.1 km^3
- VEI 5: tephra volume ≥ 1 km^3
- VEI 6: tephra volume ≥ 10 km^3
- VEI 7: tephra volume ≥ 100 km^3
- VEI 8: tephra volume ≥ 1,000 km^3

Eruption Date and C^{14} Date

These two tables are linked as they both refer to eruption date. While the Eruption Date table contains both start and end date fields, only the former is populated. End dates will be added in the future but there are very little reliable data for non-historical events and further thought is required on how to populate this entry. For historical eruptions, the day, month and year of eruption are often reported; there are currently 238 such eruptions in the LaMEVE database. Older eruptions are less precisely dated and often only the year is known. For such eruptions the dating method and associated error margin are recorded when known. As well as standard calendar years for recent eruptions, all eruption dates are recorded in years before present (BP) with 'present' taken as 1950; negative numbers in this field therefore reflect years after 1950.

For prehistoric events a number of dating techniques can be employed. For eruptions that occurred in the last ~50,000 years radiocarbon (C^{14}) dating is the most common dating method; just over 550 radiocarbon dates are currently listed in the database. However, radiocarbon years are not equivalent to standard calendar years (cal BP) because of fluctuations in atmospheric C^{14} and thus a calibration must be applied to determine the actual eruption date. This calibration is sometimes done by the author of the source reference and can be reported with or without the uncorrected C^{14} date. Where reported, the original C^{14} dates and errors are recorded in the Eruption C^{14} Date table. Where no calibrated date has been given, we have applied the calibration using a publically available tool. The method used is reported in the database, but generally the Calib program (v6.1.1) by Stuiver and Reimer (1993) is used for dates <26 Ky and the Fairbanks et al. program (2005) for events up to 50,000 cal BP. If no error has been reported for the C^{14} date, an error margin of 12% of the date is entered into the calibration program, as derived from statistical analysis of those dates with reported errors. For uncalibrated dates that fall outside the range of even the Fairbanks program, and where no literature-sourced date is available, an age of 50 Ka (the limit of the Fairbanks et al. program) is entered into the Year BP field (with the standard 12% error) but is labelled as 'assumed;' as Year BP is a required field, we need to enter an approximate date for the eruption. In all cases, calibrations are reported using the 1-sigma range (68.3% confidence intervals), as 2-sigma ranges are not available using the Fairbanks tool. Data from Bryson et al. (2006) of >2,000 calibrated radiocarbon dates from the late Pleistocene and Holocene have also been cited.

Other commonly used dating methods are Argon-Argon (^{40}Ar-^{39}Ar), Potassium-Argon (K-Ar), stratigraphy and tephrochronology. The first two are similar to radiocarbon dating insofar as they provide absolute dates, with some uncertainty, whereas the other two provide an approximate date relative to bracketing stratigraphic layers which usually have been precisely dated by another method.

Rock Type Classification and Eruption Magma

These tables contain data on the geochemistry-based rock type classifications of the erupted magma for each eruption and the magma type range respectively. The codes used in the Rock Type Classification table are the same as those used by the Smithsonian (see Figure 3), and multiple codes are often entered for a single event because magma compositions can change over the course of an eruption. Each rock type is entered as either a standard or a minor (<10%) component. The magma type range is then determined and stored in the Eruption Magma table, or marked as 'unknown' if no data have been reported. If a single rock type has been entered into the Rock Type Classification table, this becomes the magma type. If a number of distinct rock types are reported for a single eruption event, a code to cover the range of these rock types (based on SiO_2 content) is entered; for example, an eruption producing both andesite (A) and rhyolite (R) would be given the magma code AR to represent a range from andesite to rhyolite (e.g. the 30,000 BP caldera collapse eruption of Aso). There are currently 22 magma type codes entered in the relevant dictionary table: the 10 used by the Smithsonian, a further 12 based on combinations of these codes to cover eruptions with a range of rock types, and the 'unknown' category. This field can be expanded in future, if required, to encompass all possible combinations of rock types.

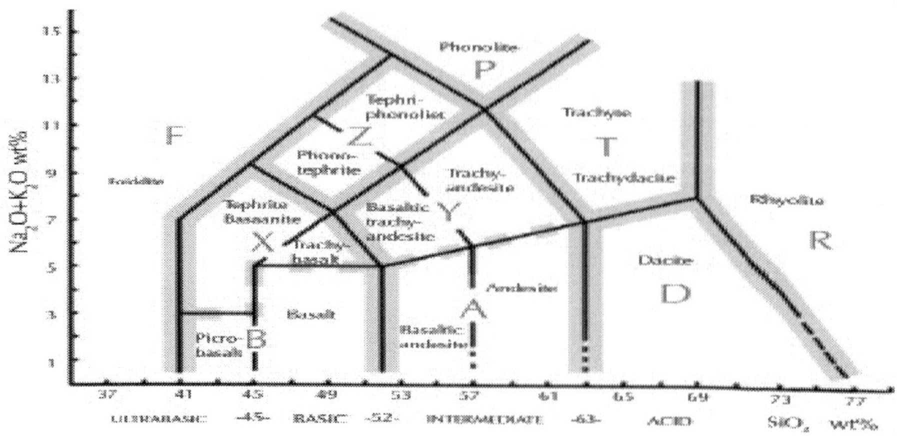

Figure 3: Codes used in *Rock Type Classification* table, taken from Siebert et al. (2010). These codes, or combinations of them, are also used in the *Eruption Magma* table.

Each magma type has an *'assumed magma density'* ranging from 2,300 kg/m^3 for rhyolitic magmas to 2,700 kg/m^3 for basaltic magmas[c]. When the magma type is unknown, a median value of 2,500 kg/m^3

is entered. These density values are used in Equation 4 to calculate the DRE volume from deposit volumes. Whilst these are relatively arbitrary values the error introduced into DRE and M values is small compared to other sources of uncertainty. For example, for a tephra volume of 100 km^3, using a density value of 2,300 kg/m^3 gives a DRE of 43.49, whereas when the density is assumed to be 2,700 kg/m^3 the DRE is 37.04. Using either of these values to calculate the magnitude would result in the same value, correct to 3 decimal places (7.000). Furthermore, assumed density values are overwritten if the literature includes a calculation of magma density, and these reported values are used instead in the DRE volume calculation.

[c]From Figure 3, the assumed densities are applied as following to each rock types: P/T/R = 2,300 kg/m^3; D = 2,400 kg/m^3; A/Y/Z = 2,500 kg/m^3; F/X/B = 2,700 kg/m^3.

Reference

An extensive table contains sources for all the LaMEVE data, and all data tables are linked to it so that the source for each piece of data can be traced. Included in the database are data from a number of pre-existing databases, such as the aforementioned Smithsonian Institution's GVP database and, more recently, a significant contribution of Japanese data following the translation of two main databases; one by the Quaternary Volcanoes in Japan2008(http://riodb02.ibase.aist.go.jp/strata/VOL_JP/EN/index.htm, last modified October 2011) and the other by Hayakawa 2010(http://gunma.zamurai.jp/database/ , last modified September 2010). The primary references for data derived from the Smithsonian's GVP will be available online from that database directly (Cottrell et al., in prep.). It also contains data gathered from a variety of published material, mostly peer-reviewed journal articles.

Commonly more than one paper is written about a particular eruption or deposit, which can result in different estimates for the same kind of data, for example an eruption date or volume. In the interests of being inclusive, all published data are included. However, a 'preferred' default value has been selected using the data quality indicators described in Tables 3 and 4, whereby the data point with the highest quality level defaults as the preferred value. If there is more than one entry with the same data quality level, the most recent publication is the default.

Searching the Database

The database is available online through a hosted by the British Geological Survey (BGS). In addition to information regarding the project's background and its future objectives, the allows users access to the entire database via a customisable search facility. This search tool is flexible and multiple criteria can be entered. A GIS-enabled graphical tool allows users to select an initial Area of Interest (AoI), if desired. An attributes search can be performed either separately, or in addition to the area search, by entering criteria for a number of fields, currently:

- Volcano name
- Volcano type (select from list)
- Eruption start date (exact/partial date or range)
- Region (select from list)
- Eruption size (range)
- Composition (select from list)
- Bulk volume (range)
- DRE volume (range)
- Column height (range)
- Tephra fall volume (range)

The relevant records are displayed in a summary table with a map of the selected area with the volcano locations indicated (see Figure 4 for an example). Volcanoes are marked on the map with small red dots with the current page of results highlighted with dark blue dots. By hovering over a blue dot it turns light blue and highlights the relevant results in the summary table. This also works in reverse. The search criteria used to generate the results are also listed.

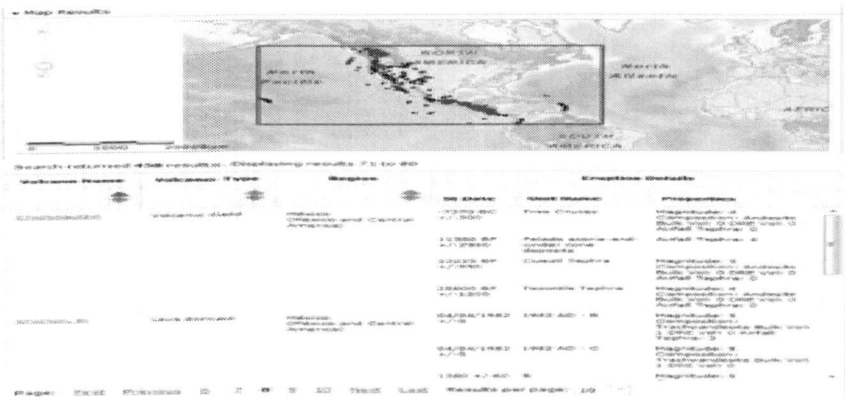

Figure 4: Example output from VOGRIPA after, in this case, using the 'Area of Interest' search tool.

By default results are ordered alphabetically by volcano name but there is also the option to sort by volcano type or region. The summary dataset can be downloaded as a spreadsheet or the search can be refined further via two methods. By selecting 'Refine Search' from a menu on the left, the original search screen is redisplayed and changes can be made. Alternatively, the 'Filter' tool can be used to select the volcano types and or regions of interest.

When an individual volcano is selected from the initial results page, a volcano summary, which includes a location map, is displayed at the top of the page (this is in addition to the full search results listed). The volcano-specific summary includes the following information:

- Volcano name
- Alternate name
- Region/Sub region
- VNUM (hyperlinked to the Smithsonian's)
- Volcano type
- Latitude/Longitude

The 'Eruption Details' section (see Figure 5) lists all the available data in the database for the selected eruptive events. Each piece of data is appended with 'literature' or 'calculated' where appropriate and details of the literature source are available by clicking on the adjacent 'Reference' tab.

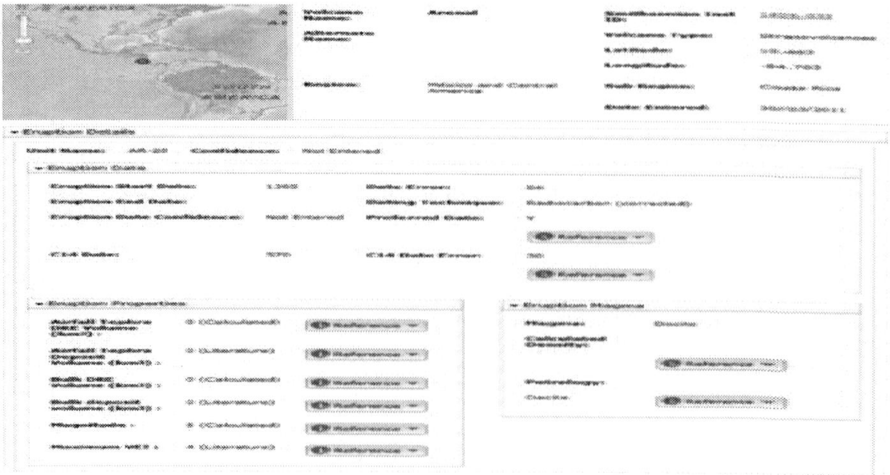

Figure 5: Example results display from VOGRIPA after an individual volcano has been selected.

DISCUSSION

Uses of the Database

LaMEVE constitutes an extensive volcanic data repository available for researchers who would have previously had to do substantial literature searches to find information on past activity. It is anticipated that civil authorities responsible for crisis management will be interested in the data to inform future management of regional volcanoes.

Whilst the database currently holds a large number of records, it is inevitably incomplete and likely contains errors. The key to improving the database will be user input and future updates as new data are published, relevant existing data are identified and existing records are updated, corrected or modified. A notable advantage to having the database online is that it will allow users to easily access the data, identify and fill gaps and correct mistakes. Any changes or updates suggested by external users will be assessed by the database manager before being entered into the database.

The database will also be used as a research tool to analyse global, regional and local patterns of volcanic activity to identify locations at high risk and gaps in knowledge about hazard and risk, as per the goals of VOGRIPA and GVM. Furthermore, the recurrence rates of different magnitude eruptions and particular hazardous phenomena, as well as the stationarity of global volcanism can be evaluated. Under-recording of events, especially those of smaller magnitude, is a significant problem (Deligne et al., 2010; Furlan, 2010) and this can be investigated with the data. A preliminary analysis of the LaMEVE database looking at patterns in the data along with an assessment of under-recording, both temporally and spatially, will be presented in a forthcoming paper.

Benefits

LaMEVE provides standardised, internally consistent data on global explosive volcanism. The database has considerable flexibility for the user who can, for example, make decisions on what data to use in analysis, e.g. decide which is the best date to use, how to combine multiple dates or whether to discard dates. We note that it is the responsibility of the user to explain and justify data usage.

LaMEVE can be queried for data that can feed into hazard and risk assessments allowing users to analyse hazard and risk within local, regional and global contexts; this should be particularly beneficial in under-studied areas. An example of an application of volcanic hazard and risk assessment is a study of 22 high priority GFDRR (Global Facility for Disaster Reduction and Recovery, World Bank) countries (Aspinall et al., 2011). Here the LaMEVE database has been used to calculate regional recurrence rates of explosive eruptions as a function of magnitude, which will be discussed in a later paper.

Planned Future Development

The LaMEVE database is the first output of VOGRIPA. A number of databases will be developed to link with the existing structure to provide data specifically relating to the main volcanic hazards (e.g. pyroclastic flows, tephra fall, lahars, debris avalanches, ballistics). These databases will be cross-linked. This hazard-specific information will enable users to determine which hazards are most common at each volcano, based on past activity, and also the impacted spatial extent.

In order to translate hazard into risk simple measures of vulnerability will be added, including population densities, building types, locations of critical infrastructure and level of monitoring. This will enable multi-scale risk mapping to identify those areas at greatest risk from future volcanic activity and so highlight where future efforts should be focussed.

The LaMEVE database is currently based largely on historical information and terrestrial geological records of Quaternary volcanoes. However, there is the potential to add information from tephra deposits in marine sediments and from indications of explosive volcanism in ice cores (e.g. from sulphate anomalies). Challenges in incorporating these data include the estimation of eruption magnitude and determination of the source volcano.

The LaMEVE database is to be made publically accessible at the time this paper is published as v1.0. For Holocene eruptions, LaMEVE data not already included in the GVP database will be imported by GVP where it will be maintained and updated in collaboration with VOGRIPA and GVM. The GVP database has been restructured for this purpose and will be available for search and download. This partnership prevents divergence and redundancy, and the division of labour will accelerate progress on both databases. The Pleistocene eruptions will be subsequently regularly updated by VOGRIPA. The current plan calls for annual updates (i.e. v2.0 released in 2013) but if other updates are required, they will be denoted as sub-versions (e.g. v1.1). There will be three mechanisms of adding new data, updating existing entries and correcting errors. Firstly, as VOGRIPA and GVM are funded until the end of 2014, GVM researchers will continue to search for and incorporate new data and make corrections when errors are identified. Secondly, researchers can send new data and modifications and draw attention to errors by contacting the database manager identified on the. Thirdly, researchers can enter data after being given clearance as a GVM user. This data will still be checked for validity and consistency by the database manager. Only the current version of the database will be available to the public; the evolving version will not be made public prior to specified release dates when it will replace the earlier version.

Difficulties

The principal difficulty in creating a database with such a wide scope is compiling information. There are two main issues. First, the data may not actually exist, either due to a lack of geological investigations or because data are unpublished. Second, not all published materials are easily accessible (either because of journal subscription or language issues). There are also potentially relevant data published in the 'grey literature,' which may also be inaccessible to academics. These concerns emphasise the value of opening the database to additional scientific contributions as undoubtedly more data exist than have been found thus far.

It is evident from even a basic analysis of the data in the database that under-recording of eruptions over time, particularly those of lower magnitudes, is a considerable problem. This will be discussed in detail in a forthcoming companion paper, and also in Deligne et al. (2010) using just the Holocene record.

Comparisons between the data are complicated by the fact that there are numerous data types as well as a mix of collection methods. For example, eruption dates can be historical or geological, which have notably different accuracies. Furthermore, a number of methods are used to determine certain eruption characteristics, such as eruption size, which complicates comparisons. Differences in data accuracy are unavoidable, particularly over the timescale covered in the database; it is the responsibility of users to take this into consideration.

Any database of this kind will, and should, evolve as more data become available and errors are corrected. Studies which use this database will need to be explicit about which version was used. Every published database will be archived to allow comparison of analyses. The development, updating and archiving of databases will require considerable effort and resources and should be sustained.

CONCLUSIONS

The LaMEVE database contains nearly 3,000 Quaternary volcanoes and just under 2,000 explosive eruptions with magnitude ≥4 spanning the last 1.8 My. It constitutes the first part of the larger VOGRIPA database and is accessible online at http://www.bgs.ac.uk/vogripa . The most common types of dates stored are historical events (238) and radiocarbon dates (558). An analysis of patterns, relationships and under-recording issues in the dataset will be published separately.

The database constitutes a major resource for those conducting research or looking for information on large magnitude explosive eruptions from around the world as it synthesises data from a wide range of sources. It is available via an easy-to-use web interface with the facility to download data. The database will be maintained and updated within the GVM project with opportunities for the volcanological community to add new data and identify errors and make corrections. The design permits future expansion, so that hazard-specific information can be stored as well as vulnerability data, which will enable global risk assessments to be carried out.

AUTHORS' CONTRIBUTIONS

HSC redesigned the database structure, undertook data quality assessment, coordinated the project and drafted the manuscript. BA provided database design expertise and assistance with the transfer of data to Oracle. SKB undertook data quality assessment and edits. EC provided data and significant revision of the manuscript. NID, NOG and LH completed initial data collection. NID also commented on the manuscript and NOG designed the original database architecture. KK translated and incorporated additional Japanese data. SCL provided advice in database requirements. JL provided expertise in GIS, created the and online access to the database. MN provided database design expertise and converted the database to Oracle. LS provided data and advice on database design in addition to comments on the draft manuscript. RSJS conceived of the study, coordinated its early development and helped draft the manuscript. ST provided data and support to KK on Japanese eruptions. EV provided data and discussions on database design. All authors read and approved the final manuscript.

ACKNOWLEDGEMENTS

The authors would like to acknowledge the funding bodies for this project: the European Research Council (VOLDIES grant), the Natural Environment Research Council (Global Volcano Model grant), the British Geological Survey and also Munich Re in the initial stages. We also thank Hayley Dunning for her input in the early data-collection stages of LaMEVE as well as work done by summer students Michael Hodge and Lucy Harper.

REFERENCES

1. Aspinall W, Auker M, Hincks TK, Mahony S, Nadim F, Pooley J, Sparks RSJ, Syre E (2011) Volcano hazard and exposure in GDRFF priority countries and risk mitigation measures - GFDRR Volcano Risk Study. Bristol University Cabot Institute and NGI Norway for the World Bank, Washington DC. NGI Report 20100806; 309pp, 3 May 2011
2. Bryson RU, Bryson RA, Ruter A (2006) A calibrated radiocarbon database of late Quaternary volcanic eruptions. Earth Discussions 1(2):123-134
3. Burden RE, Phillips JCC, Hincks TK (2011) Estimating volcanic plume heights from depositional clast size. J Geophys Res 116:B11206
4. Carey S, Sigurdsson H (1989) The intensity of plinian eruptions. Bull Volcanol 51(1):28-40
5. Carey S, Sparks RSJ (1986) Quantitative models of the fallout and dispersal of tephra from volcanic eruption columns. Bull Volcanol 48(2):109-125
6. Connolly T, Begg C (2005) Database systems: A practical approach to design, implementation, and management. Pearson Education Limited, Harlow.
7. Deligne NI, Coles SG, Sparks RSJ (2010) Recurrence rates of large explosive volcanic eruptions. J Geophys Res 115:B06203
8. Fairbanks RG, Mortlock RA, Chiu T-C, Cao L, Kaplan A, Guilderson TP, Fairbanks TW, Bloom AL, Grootes PM, Nadeau M-J (2005)

Radiocarbon calibration curve spanning 0 to 50,000 years BP based on paired 230Th/234U/238U and 14C dates on pristine corals. Quat Sci Rev 24(16–17):1781-1796

9. Fierstein J, Nathenson M (1992) Another look at the calculation of fallout tephra volumes. Bull Volcanol 54(2):156-167
10. Furlan C (2010) Extreme value methods for modelling historical series of large volcanic magnitudes. Stat Model 10(2):113-132
11. Quaternary Volcanoes in Japan (2008) Geological Survey of Japan. http://riodb02.ibase.aist.go.jp/strata/VOL_JP/EN/index.htm . Accessed regularly
12. Hayakawa Y (2010) Hayakawa's 2000-year and one million-year database. http://gunma.zamurai.jp/database . Accessed regularly
13. Mason BG, Pyle DM, Oppenheimer C (2004) The size and frequency of the largest explosive eruptions on Earth. Bull Volcanol 66(8):735-748
14. Newhall CG, Self S (1982) The Volcanic Explosivity Index (VEI): an estimate of explosive magnitude for historical volcanism. J Geophys Res 87(C2):1231-1238
15. Powell G (2006) Beginning database design. Wiley, Indianapolis.
16. Pyle DM (2000) Sizes of volcanic eruptions. In: Sigurdsson H, Houghton BF, McNutt SR, Rymer H, Stix J (eds) Encyclopedia of Volcanoes, Academic Press, London.
17. Siebert L, Simkin T, Kimberly P (2010) Volcanoes of the world. University of California Press, Berkeley.
18. Sparks RSJ, Bursik MI, Carey SN, Gilbert JS, Glaze LS, Siggurdsson H, Woods AW (eds) (1997) Volcanic plumes Wiley, New York.
19. Stuiver M, Reimer PJ (1993) Extended 14C data base and revised CALIB 3.0 14C age calibration program. Radiocarbon 35(1):215-230
20. Wilson L, Sparks RSJ, Huang TC, Watkins ND (1978) The control of volcanic column heights by eruption energetics and dynamics. J Geophys Res 83(B4):1829-1836

Chapter 8

Impact of additional module training on the level of basic life support knowledge of first year students at the University of Maribor

Damjan Lešnik[1], Bojan Lešnik[1], Jerneja Golub[2], Miljenko Križmarić[2, 3], Štefan Mally[2, 4] and Štefek Grmec[1, 2, 3, 5]

[1]Center for Emergency Medicine, Ulica talcev 9, 2000 Maribor, Slovenia

[2]Faculty of Medicine, University of Maribor, Slomškov trg 15, 2000 Maribor, Slovenia

[3]Faculty of Health Sciences,Žitna ulica 15, 2000 Maribor; Slovenia

[4]Center for Emergency Medicine, Ulica talcev 9, 2000 Maribor, Slovenia

[5]Faculty of Medicine, University of Ljubljana, Vrazov trg 2, 1104 Ljubljana, Slovenia

ABSTRACT

Aim

The aim of this study was to investigate the impact of additional (two versus one session) basic life support (BLS) training of university students on knowledge and attitude concerning the performance of cardiopulmonary resuscitation.

Methods

A total of 439 students in three separate groups were tested: those with no prior BLS training; BLS training in high school (part of the driver's education course); and BLS training in high school (in the driver's education course) and additional BLS training at the university.

Results

Our study showed the best results of BLS education in a group of university students who took an additional BLS module approximately half a year after the driver's education BLS course. In our study we observed equal levels of knowledge between the group with BLS training in high school and the group without any formal BLS education. The questionnaire revealed a disappointing level of knowledge about BLS in both groups.

Conclusion

Additional basic life support training (two BLS training sessions: high school and university) improves retention of knowledge and attitudes concerning performing CPR in first year university students.

INTRODUCTION

Recent studies have emphasized that bystander cardiopulmonary resuscitation (CPR) is a very important contributing factor in the survival of out-of-hospital cardiac arrest (OHCA) patients [1-10]. The practice of basic life support (BLS) by lay people is therefore essential for sufficient functioning of the chain of survival and is definitely an important part of effective emergency services for a patient needing resuscitation [11]. However, lay people can only play their role within the chain of survival if they are adequately trained and if continuous repetitions of relevant training information are offered and used [12]. Introducing CPR training in high school and university settings has been widely recommended as a long-term strategy to educate the wider community. In general, students have poor theoretical knowledge, although most of them are willing and motivated to learn CPR [12-17]. A pyramidal teaching approach involving students who had BLS training shows potential for spreading BLS knowledge to lay people [18]. In a previous study we confirmed that the potential bystander in our community is generally poorly educated about performing CPR, but willing to gain knowledge and skills in BLS and to follow dispatchers' instructions [19].

The aim of this study was to investigate the impact of additional basic life support (BLS) training of university students on knowledge and attitude for performing cardiopulmonary resuscitation (one training session vs. two training sessions).

METHODS

The study was conducted in the context of the campaign program "Education of lay people in BLS in the Maribor area" and arranged by the authors. The campaign program was designed to facilitate the wider dissemination of BLS skills and knowledge in the local population.

Data for the study were collected in the spring of 2009. We compared the knowledge of BLS in three university student groups at the end of the first year of college: those with no prior BLS training, those who had BLS training in high school (in the driver's education course) approximately 13 months before testing, and those who had BLS training in high school and additional BLS training at the university (within 6 months after the driver's education BLS course and approximately 8 months before testing). Approximately 90% of high school students successfully accomplished the driver's education course. The 10% of high school students who failed the driver's education course (and thus did not have any other opportunity for BLS training in high school) would therefore be expected to perform poorly in the BLS program at the university level. Participants in the first two groups were students of the Faculty of Electrical Engineering/Computer Science and of the Faculty of Education. In the third group were students of the Faculty of Medicine and Faculty of Health Sciences.

The driver's education BLS courses were guided by health workers who are first aid instructors certified by the Slovenian Red Cross organization and supervised by the Slovenian Resuscitation Council (pyramid teaching methods). The driver's education BLS training is part of the first aid course for all driver's license candidates. The BLS training includes a lecture (1 h) and practical training in small groups (4 h) in accordance with European Resuscitation Council (ERC) recommendations. The additional BLS module was given by ERC instructors. The driver's education BLS training and the BLS module have identical structures, presentations, contents, tools and equipment. Both groups were trained by ERC instructors. The Red Cross group had ERC licenses for the BLS course, and the ERC group had licenses for the BLS and Advanced Life Support course.

Knowledge of BLS was tested with a questionnaire in accordance with the 2005 ERC guidelines for BLS and approved by the deans of all four faculties, who were informed about the results at the end of the study and were advised of a concrete plan of action to improve the BLS knowledge of their students. According to the Declaration of Helsinki, data were made anonymous [20]. A standardized questionnaire with 28 items (see Tables 1, 2 and 3) included checkboxes or open answer areas, and contained information about the intention of the survey. It was presented to students and collected personally by the authors at the end of testing in hard form.

Table 1: Comparison between the groups with a driver's education BLS course followed by BLS module training and the group with the driver's education BLS course only

Question (correct answer)	BLS driver's education course + BLS module course	BLS driver's education course only	p-value
	N = 197	N = 179	
Gender [Male/all]	Male: 45/197 (23%)	Male: 65/179 (36%)	0.004
Age [years ± SD]	19.6 ± 0.7 (18-22)	19.7 ± 0.9 (19-24)	NS
Are you living in an urban area?	Yes: 69/197 (35%)	Yes: 57/179 (32%)	NS
Would you recognize a situation in which CPR is needed? §	Mean ± SD: 4.4 ± 0.6	Mean ± SD: 3.9 ± 0.7	<0.001
In your opinion, how well do you know BLS? §§	Mean ± SD: 3.6 ± 0.6	Mean ± SD: 2.9 ± 0.6	<0.001
In your opinion, how effectively would you perform CPR? §§	Mean ± SD: 3.4 ± 0.5	Mean ± SD: 2.8 ± 0.6	<0.001
Would you perform rescue breathing in everyone when needed?	Yes: 197/197 (100%)	Yes: 172/179 (96%)	0.005
What is the emergency telephone number? (112)	Correct answer: 196/197 (99%)	Correct answer: 157/179 (88%)	<0.001
How long should a lay person perform CPR? (until the rescue unit arrives)	Correct answer: 89/197 (45%)	Correct answer: 28/179 (16%)	<0.001
Have you ever done CPR?	Yes: 20/197 (10%)	Yes: 13/179 (7%)	0.323
Have you ever seen someone to lose consciousness?	Yes: 74/197 (38%)	Yes: 68/179 (38%)	0.932

Would you be willing to listen to dispatcher's instructions and perform CPR untill the rescue unit arrives?	Yes: 107/197 (54%)	Yes: 119/179 (67%)	0.016
Would you do CPR with no scruples about it, when necessary?	Yes: 84/197 (43%)	Yes: 28/179 (16%)	<0.001
Did you get the most of your knowledge of CPR at BLS driver course?	Yes: 105/197 (53%)	Yes: 160/179 (89%)	<0.001
Are you interested in gaining more skills and knowlege of BLS?	Yes: 191/197 (97%)	Yes: 165/179 (92%)	0.039
Would you be willing to gain additional BLS knowlege in a form of BLS courses?	Yes: 154/197 (78%)	Yes: 107/179 (60%)	<0.001
What is the correct position of hands when performing compressions? (on sternum in the middle of the chest)	Correct answer: 186/197 (94%)	Correct answer: 139/179 (78%)	<0.001
What is CPR ratio between chest compressions and breathing in adults? (30:2)	Correct answer: 189/197 (189%)	Correct answer: 70/179 (39%)	<0.001
Frequency of chest compressions in adults? (100/min)	Correct answer: 157/197 (80%)	Correct answer: 33/179 (18%)	<0.001
The depth of chest compressions for effective CPR is? (4-5 cm)	Correct answer: 150/197 (76%)	Correct answer: 61/179 (34%)	<0.001
What is Heimlich sign? (grasping one`s own throat unable to breath)	Correct answer: 151/197 (77%)	Correct answer: 51/179 (29%)	<0.001
How would you do rescue breathing in mouth to mouth resuscitation? (blow steadily about 1s as in normal breathing)	Correct answer: 83/197 (42%)	Correct answer: 21/179 (12%)	<0.001

What is important for effective CPR in adults? (to do chest compessions together with rescue breathing)	Correct answer: 140/197 (71%)	Correct answer: 143/179 (80%)	0.048
What is the purpose of automatic defibrillator (AED)? (to end some of lifethreating heart rhythm disturbances)	Correct answer: 129/197 (66%)	Correct answer: 79/179 (44%)	<0.001
What is the sequence of CPR when using AED? (CPR is performed as usual, additionally placing the electrodes and following the AED instruction)	Correct answer: 83/197 (42%)	Correct answer: 64/179 (36%)	0.206

§Five-point Likert scale: 5- always, 1- never.

§§Five-point Likert scale: 5-excellent, 1-extremely poor.

Lešnik et al. International Journal of Emergency Medicine 2011 4:16, doi: 10.1186/1865-1380-4-16

Table 2: Comparison between the group with no BLS training and the group with the driver's education BLS course only

Question (correct answer)	No BLS training	BLS driver's education course only	p-value
	N = 63	N = 179	
Gender [Male/Female]	Male: 9/63 (14%)	Male: 65/179 (36%)	<0.001
Age [years ± SD]	19.1 ± 0.8 (18-21)	19.7 ± 0.9 (19-24)	<0.001
Are you living in an urban area?	Yes: 19/63 (30%)	Yes: 57/179 (32%)	0.804
Would you recognize a situation in which CPR is needed? §	Mean ± SD: 3.5 ± 0.5	Mean ± SD: 3.9 ± 0.7	<0.001
In your opinion, how well do you know BLS? §§	Mean ± SD: 3.6 ± 0.6	Mean ± SD: 2.9 ± 0.6	<0.001

In your opinion, how effectively would you perform CPR? §§	Mean ± SD: 2.6 ± 0.6	Mean ± SD: 2.8 ± 0.6	0.024
Would you perform rescue breathing in everyone when needed?	Yes: 46/63 (73%)	Yes: 172/179 (96%)	<0.001
What is the emergency telephone number? (112)	Yes: 58/63 (92%)	Yes: 157/179 (88%)	0.345
How long should a lay person perform CPR? (until the rescue unit arrives)	Correct answer: 8/63 (13%)	Correct answer: 28/179 (16%)	0.572
Have you ever done CPR?	Yes: 2/63 (3%)	Yes: 13/179 (7%)	0.247
Have you ever seen someone to lose consciousness?	Yes: 22/63 (35%)	Yes: 68/179 (38%)	0.665
Would you be willing to listen to dispatcher's instructions and perform CPR untill the rescue unit arrives?	Yes: 58/63 (92%)	Yes: 119/179 (67%)	<0.001
Would you do CPR with no scruples about it, when necessary?	Yes: 7/63 (11%)	Yes: 28/179 (16%)	0.3792
Are you interested in gaining more skills and knowlege of BLS?	Yes: 63/63 (100%)	Yes: 165/179 (92%)	0.0222
Would you be willing to gain additional BLS knowlege in a form of BLS courses?	Yes: 61/63 (97%)	Yes: 107/179 (60%)	<0.001
What is the correct position of hands when performing compressions? (on sternum in the middle of the chest)	Correct answer: 48/63 (76%)	Correct answer: 139/179 (78%)	0.812
What is CPR ratio between chest compressions and breathing in adults? (30:2)	Correct answer: 23/63 (37%)	Correct answer: 70/179 (39%)	0.715
Frequency of chest compressions in adults? (100/min)	Correct answer: 10/63 (16%)	Correct answer: 33/179 (18%)	0.647

The depth of chest compressions for effective CPR is? (4-5 cm)	Correct answer: 21/63 (33%)	Correct answer: 61/179 (34%)	0.914
What is Heimlich sign? (grasping one`s own throat unable to breath)	Correct answer: 18/63 (29%)	Correct answer: 51/179 (29%)	0.990
How would you do rescue breathing in mouth to mouth resuscitation? (blow steadily about 1s as in normal breathing)	Correct answer: 8/63 (13%)	Correct answer: 21/179 (12%)	0.839
What is important for effective CPR in adults? (to do chest compessions together with rescue breathing)	Correct answer: 49/63 (78%)	Correct answer: 143/179 (80%)	0.722
What is the purpose of automatic defibrillator (AED)? (to end some of lifethreating heart rhythm disturbances)	Correct answer: 27/63 (43%)	Correct answer: 79/179 (44%)	0.861
What is the sequence of CPR when using AED? (CPR is performed as usual, additionally placing the electrodes and following the AED instruction)	Correct answer: 21/63 (33%)	Correct answer: 64/179 (36%)	0.729

§Five-point Likert scale: 5- always, 1- never.

§§Five-point Likert scale: 5-excellent, 1-extremely poor.

Lešnik et al. International Journal of Emergency Medicine 2011 4:16, doi: 10.1186/1865-1380-4-16

Table 3: Comparison between the group of the Faculty of Health Sciences and the group of the Faculty of Medicine

Question (correct answer)	Faculty of Health Sciences	Faculty of Medicine	p-value
	N = 97	N = 100	
Gender [Male/Female]	Male: 10/97 (10%)	Male: 35/100 (35%)	<0.001
Age [years ± SD]	19.7 ± 0.7 (19-22)	19.5 ± 0.7 (18-22)	0.914
Are you living in an urban area?	Yes: 28/97 (29%)	Yes: 41/100 (41%)	0.074
Would you recognize a situation in which CPR is needed? §	Mean ± SD: 4.4 ± 0.5	Mean ± SD: 4.4 ± 0.6	0.685
In your opinion, how well do you know BLS? §§	Mean ± SD: 3.6 ± 0.6	Mean ± SD: 3.6 ± 0.6	0.928
In your opinion, how effectively would you perform CPR? §§	Mean ± SD: 3.4 ± 0.5	Mean ± SD: 3.4 ± 0.6	0.864
Would you perform rescue breathing in everyone when needed?	Yes: 97/97 (100%)	Yes: 100/100 (100%)	-
What is the emergency telephone number? (112)	Yes: 96/97 (99%)	Yes: 100/100 (100%)	0.309
How long should a lay person perform CPR? (until the rescue unit arrives)	Correct answer: 43/97 (44%)	Correct answer: 46/100 (46%)	0.814
Have you ever done CPR?	Yes: 11/97 (11%)	Yes: 9/100 (9%)	0.587
Have you ever seen someone to lose consciousness?	Yes: 42/97 (43%)	Yes: 32/100 (32%)	0.102
Would you be willing to listen to dispatcher's instructions and perform CPR untill the rescue unit arrives?	Yes: 66/97 (68%)	Yes: 41/100 (41%)	<0.001

Would you do CPR with no scruples about it, when necessary?	Yes: 38/97 (39%)	Yes: 46/100 (46%)	0.333
Did you get the most of your knowledge of CPR at BLS driver course?	Yes: 54/97 (56%)	Yes: 51/100 (51%)	0.511
Are you interested in gaining more skills and knowlege of BLS?	Yes: 94/97 (97%)	Yes: 97/100 (97%)	0.970
Would you be willing to gain additional BLS knowlege in a form of BLS courses?	Yes: 79/97 (81%)	Yes: 75/100 (75%)	0.274
What is the correct position of hands when performing compressions? (on sternum in the middle of the chest)	Correct answer: 90/97 (93%)	Correct answer: 96/100 (96%)	0,326
What is CPR ratio between chest compressions and breathing in adults? (30:2)	Correct answer: 94/97 (97%)	Correct answer: 95/100 (95%)	0.498
Frequency of chest compressions in adults? (100/min)	Correct answer: 77/97 (79%)	Correct answer: 80/100 (80%)	0.914
The depth of chest compressions for effective CPR is? (4-5 cm)	Correct answer: 76/97 (78%)	Correct answer: 74/100 (74%)	0.474
What is Heimlich sign? (grasping one`s own throat unable to breath)	Correct answer: 83/97 (86%)	Correct answer: 68/100 (68%)	0.004
How would you do rescue breathing in mouth to mouth resuscitation? (blow steadily about 1s as in normal breathing)	Correct answer: 33/97 (34%)	Correct answer: 50/100 (50%)	0.023
What is important for effective CPR in adults? (to do chest compessions together with rescue breathing)	Correct answer: 61/97 (63%)	Correct answer: 79/100 (79%)	0.013

What is the purpose of automatic defibrillator (AED)? (to end some of lifethreating heart rhythm disturbances)	Correct answer: 47/97 (48%)	Correct answer: 82/100 (82%)	<0.001
What is the sequence of CPR when using AED? (CPR is performed as usual, additionally placing the electrodes and following the AED instruction)	Correct answer: 39/97 (40%)	Correct answer: 44/100 (44%)	0.590

§Five-point Likert scale: 5- always, 1- never.

§§Five-point Likert scale: 5-excellent, 1-extremely poor.

Lešnik et al. International Journal of Emergency Medicine 2011 4:16, doi: 10.1186/1865-1380-4-16

Comparisons were made among the groups and within the third group with the BLS module course, where the collected data were compared between the two faculties (Faculty of Medicine and Faculty of Health Sciences). Statistical analyses were performed using SPSS for Windows, release 12.0; SPSS, Chicago IL. Wilcoxon signed rank test, t-test and Fisher's exact test were used where appropriate. Descriptive values of variables were expressed as average, standard deviation and percentages. Power analysis was made by using G-Power™ 3.0.10 for Microsoft Windows XP™ (Microsoft Inc., Redmond, WA). Wilcoxon signed rank test, t-test and Fisher's exact test were used where appropriate. For data not normally distributed, the Wilcoxon signed rank test was used. All *p*values of less than 0.05 were considered to indicate statistical significance.

RESULTS

A total of 439 students (118 men and 321 women; average age 19.5 +/- 0.8 years) participated in the study. There were 197 participants with additional BLS training (Faculty of Medicine and Faculty of Health Sciences), 179 participants who had taken the driver's education BLS course and 63 participants who had had no BLS training. All results are shown in Tables 1, 2 and 3.

Compared to the group who received an additional BLS training module, we found that the group with BLS training from the driver's education course was more willing to follow dispatchers' instructions by telephone to perform CPR (67% vs. 54%; $p < 0.05$) and less willing to take BLS training, especially in course form ($p < 0.001$) (Table 1). Only 16 percent of students with BLS training from the driver's education course were prepared to start CPR without any delay when necessary compared to 43 percent in the group that had received the BLS module ($p < 0.001$). On the five-point Likert scale, we found higher results of self-assurance in one's own knowledge in the group with the BLS module ($p < 0.001$). The same group was better informed about the emergency number in Slovenia ($p < 0.001$). There was also a significantly higher rate of correct answers about closed chest compression, breathing/ventilation, recognizing the Heimlich sign, automatic defibrillator use and how to approach unconscious victims in the BLS module group ($p < 0.001$).

When comparing the groups with and without the driver's education BLS course (Table 2), we found no statistical differences in knowledge about closed chest compression, breathing/ventilation, the Heimlich sign, automatic defibrillator use and how to approach unconscious victims. The group without a BLS course was more willing to follow dispatchers' instructions by telephone to perform CPR (92% vs. 67%; $p < 0.001$) and to take part in a BLS training course (97% vs. 60%; $p < 0.001$). The same group showed less self-assurance in their own knowledge of BLS (measured by five-point Likert scale; $p < 0.001$).

Within the group with the BLS module, we compared the students of the Faculty of Medicine with students of the Faculty of Health Sciences. Medical students have better knowledge of ventilation ($p = 0.02$), automatic defibrillator use ($p < 0.001$) and coordination in CPR ($p = 0.01$). The students of the Faculty for Health Sciences were more willing to follow dispatchers' instructions by telephone ($p < 0.001$) and recognized the Heimlich sign better ($p = 0.004$).

DISCUSSION

Since previous studies have found that university students showed poor theoretical knowledge and demonstrated willingness and motivation for courses on BLS [12-17], we sought to examine characteristics of students of the University of Maribor. Our survey suggests that an additional module of BLS training in the first academic year improves theoretical knowledge in students and their preparedness to perform CPR. This study also highlights some notable differences between the two faculties inside the group of students who received an additional BLS module.

Perkins et al. [21] found that care of the acutely ill patient in the hospital is often suboptimal. Poor recognition of critical illness combined with a lack of knowledge and failure to appreciate clinical urgency of a situation has been identified as a contributory factor. They confirmed that the present training of medical students in these important skills is fragmented. The nominal group in this study identified 71 essential and 16 optional competencies that students should possess by graduation and proposed that these competencies should form a core set for undergraduate training in resuscitation and acute care [21]. Beckers et al. reported on the Medical Reform Curriculum Aachen, which has a 3-week interdisciplinary introduction to emergency medical care in the first semester of medical school. Besides skill training in the basics of emergency medical care (BLS, early defibrillation), practical training in other lifesaving techniques (e.g., immobilization skills) and basic principles of daily clinical care are included. The course evaluation data clearly showed acceptance of the new approach and enhanced possibilities of extending implementation of relevant topics concerning emergency medical care within the Medical Reform Curriculum Aachen [22]. Das and Elzubeir confirmed the importance of training physicians and other health care professionals in first aid and BLS in the form of formal training in the first year of medical school. In their study they found that students were uniformly enthusiastic and highly motivated by the program [23]. Self-assessed confidence in the ability to perform skills on their own after completing the program was moderately correlated with the perceived frequency of opportunity to practice many skills. There was nevertheless a consistent desire for more time to practice.

In the curriculum of first year students at the Faculty of Medicine and the Faculty of Health Sciences in Maribor, formal BLS training as a BLS module course and practical training in other lifesaving techniques have been incorporated. As we confirmed in our investigation, students with this form of BLS education have shown better theoretical knowledge and higher confidence in their own skills than students with BLS training from the driver's education course and have been more motivated for additional learning in the form of courses. As the results of our survey together with those of several other studies suggest, BLS education in a module form is very effective and could become obligatory for all university and high school students [14,15,17,24-27], or even for those in primary school [16,27,28]. Our agreements with deans of different faculties in Maribor are in the final phase, and we are close to starting a major campaign of BLS courses at our university [19, 29]. In our survey, we observed a substantial portion of students who had insufficient knowledge and skills to perform effective ventilation in CPR after BLS training. This suggests that the courses in the first year of college should be oriented to effective chest compression, safe and early defibrillation, and good communication with the dispatcher.

We have demonstrated that there are no differences in key points of BLS knowledge between students with driver's education BLS training and students without BLS training. This observation takes into consideration the quality of driver's education BLS first aid courses organized by the national Red Cross. In our opinion reevaluation of the first aid instructor education and of the form of BLS training in driver's education courses is necessary. Our results confirm that retention of CPR skills is poor. The solution for this problem is to improve instructor training and to use contemporary teaching methods (more frequent refresher courses for instructors). Also a general testing of knowledge retention with uniform criteria should be applied [30, 31]. Detailed CPR quality assessment, careful monitoring of the quality of instructions, awareness of instructors when inadequate CPR is demonstrated and feedback should be integrated into the training to ensure optimal performance in a real-life resuscitation - it is time to stop the blind leading the blind [32]. Optimal refresher training in BLS is another option for better retention of CPR skills. Refresher training intervals should not exceed 7 months [33, 34]. Maybe our two-step method (driver's education BLS course followed by BLS module training) is a practical way to improve the retention of key BLS knowledge. Traditional classroom or simulation-based learning could be combined with DVD or website self-instructional systems in order to reach and reinstruct individuals who are unwilling to participate in a live course [35-37]. Medical students of the Faculty of Medicine have founded a society called "For Life." This group of young people, under the supervision of ERC instructors and teachers, organizes BLS and first aid courses for lay persons in various places. In 2009 they successfully carried out 44 BLS courses with 1,110 participants, including school children. Their next project is a campaign in primary schools, where more medical students will participate as instructors in BLS training [38].

LIMITATIONS

Our study has some limitations. The group of students without BLS training was relatively small, so the results can only partially be transferred to a general population of students with no prior BLS training. We have compared two different levels of BLS trainers (BLS vs. BLS + ALS trainers), but both were instructed in accordance with the methodical and didactic instructions of the ERC as recommended in the 2005 ERC Guidelines. We compared BLS knowledge among three different faculties, and nurses/medical students probably had some interest and affinity compared to teachers and engineers. However, in the group with no prior BLS training we observed great willingness to gain knowledge and skills in BLS and to follow dispatchers' instructions. This information supports ideas about interest and motivation in non-medical students. The primary goal of this article is to motivate other faculty members to help with the progress in improving teaching methods in BLS training and reactivating BLS education for all university students.

CONCLUSIONS

Our study has shown the best results for BLS education in a group of university students who took an additional BLS module approximately a half year after the driver's education BLS course. The two-step method is a practical way to improve the retention of knowledge concerning BLS. We did not observe any differences in theoretical knowledge between the group of students who had had BLS training during the driver's education course and the group without any formal BLS education. The questionnaire revealed a disappointing level of knowledge of the fundamentals of BLS in both groups. However, there was a welcomed willingness of these students as potential bystanders to take BLS training and to follow dispatchers' instructions by telephone if CPR is indicated.

We must recognize this fact as an emergency call to organize BLS module courses for all university students in the first year of study.

AUTHORS' CONTRIBUTIONS

GŠ conceived of the study, and participated in its design and coordination. LD, LB and GJ participated in acquisition of data, LD and LB also contributed to concepcion of the study. KM participated in interpretation of data and performed the statistical analysis. MŠ participated in the design of the study and drafted the manuscript. All authors read and approved the final manuscript.

REFERENCES

1. Stratton SJ, Rashi P: Los Angeles County Prehospital Care Coordinators. Out-of-hospital unwitnessed cardiopulmonray collapse and no-bystander CPR: a practical addition to resuscitation termination guidelines. *J Emerg Med* 2008, 35:175-9.
2. Vadeboncoeur T, Bobrow BJ, Clark L, Kern KB, Sanders AB, Berg RA, Ewy GA: The Save Hearts in Arizona Registry and Education (SHARE) program: who is performing CPR and where are they doing it? *Resuscitation* 2007, 75:68-75. PubMed Abstract |
3. Herlitz J, Svensson L, Holmberg S, Ängquist KA, Young M: Efficacy of bystander CPR: intervention by lay people and by health care professionals. *Resuscitation* 2005, 66:291-5.
4. Nordberg P, Hollenberg J, Herlitz J, Rosenqvist M, Svensson L: Aspects on the increase in bystander CPR in Sweden and its association with outcome. *Resuscitation* 2009, 80:329-33.
5. Hollenberg J, Herlitz J, Lindqvist J, Riva G, Bohm K, Rosenqvist M, Svensson L: Improved survival after out-of-hospital cardiac arrest is associated with an increase in proportion of emergency crew-witnessed cases and bystander cardiopulmonary resuscitation. *Circulation* 2008, 1184:389-96. Publisher Full Text
6. Swor RA, Jackson RE, Compton S, Domeier R, Zalenski R, Honeycutt L, Kuhn GJ, Frederiksen S, Pascual RG: Cardiac arrest

in private locations: different strategies are needed to improve outcome. *Resuscitation* 2003, 58:171-6.

7. Swor RA, Fahoome G, Compton S: Potential impact of a targeted cardiopulmonary resuscitation program for older adults on survival from private residence cardiac arrest. *Acad Emerg Med* 2005, 12:7-12.
8. Shiraki T, Osawa K, Suzuki H, Yoshida M, Takahashi N, Takeuchi K, Tanakaya M, Kohno K, Saito D: Incidence and outcomes of out-of-hospital cardiac arrest in the eastern part of Yamaguchi prefecture. *Int Heart J* 2009, 54:489-500.
9. Iwami T, Kawamura T, Hiraide A, Berg RA, Hayashi Y, Nishiuchi T, Kajino K, Yonemoto N, Yukioka H, Sugimoto H, Kakuchi H, Sase K, Yokoyama H, Nonogi H: Effectiveness of bystander-initiated cardiac-only resuscitation for patients with out-of- hospital cardiac arrest. *Circulation* 2007, 116:2900-7.
10. Abe T, Tokuda Y, Ishimatu S, SOS-KANTO study group: Predictors for good cerebral performance among adult survivors of out-of-hospital cardiac arrest. *Resuscitation* 2009, 80:431-6.
11. Wiese CH, Wilke H, Bahr J, and Graf BM: Practical examination of bystanders performing Basic Life Support in Germany: a prospective manikin study. *BMC Emerg Med* 2008, 8:14.
12. Parnell MM, Larsen PD: Poor quality teaching in lay person CPR courses. *Resuscitation* 2007, 73:271-8.
13. Reder S, Quan L: Cardiopulmonary resuscitation training in Washington state public high schools. *Resuscitation* 2003, 56:283-8.
14. Niemi-Murola L, Mäkinen M, Castren M, ECCE Study Group: Medical and nursing students attitudes toward cardiopulmonary resuscitation and current practice guidelines. *Resuscitation* 2007, 72:257-63.
15. Hamasu S, Morimoto T, Kuramoto N, Horiguchi M, Iwami T, Nishiyama C, Takada K, Kubota Y, Seki S, Maeda Y, and Sakai Y, Hiraide A: Effects of BLS training on factors associated with attitude toward CPR in college students. Resuscitation. 2009, 80(3):359-64.
16. Miro O, Jimenez-Fabrega X, Espigol G, *et al.*: Effects of BLS training. Teaching Basic Life Support to 12-16 year olds in

Barcelona schools: views of head teachers. *Resuscitation* 2006, 70:107-16.

17. Parnell MM, Perason J, Galletly DC, Larsen PD: Knowledge of and attitudes towards resuscitation in New Zealand high-school students. *Emerg Med J* 2006, 23:899.
18. Toner P, Connolly M, Laverty L, McGrath P, Connolly D, McCluskey DR: Effects of BLS training. Teaching basic life support to school children using medical students and teachers in a "peer-training" model - results of the "ABC for life" programme. *Resuscitation* 2007, 75:169-75.
19. Grmec Š, Križmari M, Mally Š, Koželj A, Špindler M, Lešnik B: Utstein style analysis of out-of-hospital cardiac arrest-bystander CPR and end expired carbon dioxide. *Resuscitation* 2007, 72:404-14.
20. Salako SE: The Declaration of Helsinki 2000: Ethical principles and the dignity of difference. *Med Law* 2006, 25:341-54.
21. Perkins GD, Barrett H, Bullock I, Gabbott DA, Nolan JP, Mitchell S, Short A, Smith CM, Smith GB, Todd S, Bion JF: The Acute Care Undergraduate Teaching (ACUTE) Intiative; consensus development of core compentencies in acute care for undergraduates in the United Kingdom. *Intensive Care Med* 2005, 31:1627-33.
22. Beckers S, Fries M, Bickenbach J, Hoffmann N, Classen-Linke I, Killersreiter B, Wainwright U, Rossaint R, Kuhlen R: Introduction Course Working Group. Evaluation of a new approach to implement structured, evidence-based emergency medical care in undergraduate medical education in Germany. *Resuscitation* 2005, 65:345-56.
23. Das M, Elzubeir M: First aid and basic life support skills training early in medical curriculum: ciriculum issues, outcomes, and confidence of students. *Teach Learn Med* 2001, 13:240-6.
24. Hubble MW, Bachman M, Price R, Martin N, Huie D: Willingness of high school students to perform cardiopulmonary resuscitation and automated external defibrillation. *Prehosp Emerg Care* 2003, 7:219-24.
25. Al-Turki YA, Al-Fraih YS, Jalaly JB, *et al.*: Knowledge and attitudes towards cardiopulmonary resuscitation among university students in Riyadh, Saudi Arabia. *Saudi Med J* 2008, 29:1306-9.

26. Sosada K, Zurawinski W, Stepien T, Makarska J, Myrcik D: Evaluation of the knowledge of teachers and high school students in Silesia on the principals of first aid. *Wiad Lek* 2002, 55:883-9.
27. Reder S, Quan L: Cardiopulmonary resuscitation training in Washington state public schools. *Resuscitation* 2003, 56:283-8.
28. Baser M, Coban S, Tasci S, Sungar G, Bayat M: Evaluating first-aid knowledge and attitude of a sample of Turkish primary school teachers. *J Emerg Nurs* 2007, 33:428-32.
29. Grmec Š, Mally Š: Emergency medicine in Slovenia-Emergency center, prehospital emergency medicine and academic emergency medicine. *Lije Vjesn* 2009, 31(Suppl 1):16-20.
30. Kaye W, Rallis SF, and Mancini ME: The problem of poor retention of cardiopulmonary resuscitation skills may lie with the instructor, not the learner or the curriculum. *Resuscitation* 1991, 21:67-87.
31. Hazinski MF, Markenson D, Neish S, Gerardi M, Hootman J, Nichol G, Taras H, Hickey R, Oconnor R, Potts J, van der Jagt E, Berger S, Schexnayder S, Garson AJr, Doherty A, Smith S: Response to cardiac arrest and selected life-threatening medical emergencies: The medical emergency response plan for schools--a statement for healthcare providers, policymakers, school administrators, and community leaders. *Ann Emerg Med* 2004, 43:83-99.
32. Arshid M, Lo TY, and Reynolds F: Quality of cardiopulmonary resuscitation (CPR) during pediatric resuscitation training: time to stop the blind leading the blind. *Resuscitation* 2009, 80:558-60.
33. Woollard M, Whitfield R, Newcombe RG, Colquhoun M, Vetter N, and Chamberlain D: Optimal refresher training intervals for AED and CPR skills: a randomised controlled trial. *Resuscitation* 2006, 71:237-47.
34. Chamberlain D, Smith A, Woollard M, Colquhoun M, Handley AJ, Leaves S, Kern KB: Trials of teaching methods in basic life support (3): comparison of simulated CPR performance after first training and at 6 months, with a note on the value of re-training. *Resuscitation* 2002, 53:179-87.
35. Braslow A, Brennan RT, Newman MM, Bircher NG, Batcheller AM, Kaye W: CPR training without an instructor: development and evaluation of a video self-instructional system for effective

performance of cardiopulmonary resuscitation. *Resuscitation* 1997, 34:207–20.

36. Einspruch EL, Lynch B, Aufderheide TP, Nichol G, and Becker L: Retention of CPR skills learned in a tradiotional AHA Heartsaver course versus 30-min video self-training: a controlled randomized study. *Resuscitation* 2007, 74:476-86.
37. Bjorshol CA, Lindner TW, Soreide E, and Moen L, Sunde K: Hospital employees improve basic life support skills and confidence with a personal resuscitation manikin and 24-min video instruction. *Resuscitation* 2009, 80:898-902.
38. Breckwoldt J, Beetz D, Schnitzer L, Waskow C, Arntz HR, Weimann J: Medical students teaching basic life support to school children as a required element of medical education-a randomised controlled study comparing three different approaches to fifth year medical training in emergency medicine. *Resuscitation* 2007, 74:158–65.

Citations

CHAPTER 1

Dara Nix-Stevenson, Human Response to Natural Disasters, DOI: 10.1177/2158244013489684

CHAPTER 2

Nicole K Zagelbaum, Kevin C Heslin, Judith A Stein, Josef Ruzek, Robert E Smith, Tam , Nyugen and Aram Dobalian, Factors influencing readiness to deploy in disaster response: findings from a cross-sectional survey of the Department of Veterans Affairs Disaster Emergency Medical Personnel System doi:10.1186/1471-227X-14-16.

CHAPTER 3

Edbert B Hsu, Tamara L Thomas, Eric B Bass, Dianne Whyne, Gabor D Kelen, and Gary B Green, Healthcare Worker Competencies for Disaster Training, doi:10.1186/1472-6920-6-19.

CHAPTER 4

James W Hansen, Stephen Zebiak, and Kevin Coffey, Shaping Global Agendas on Climate Risk Management and Climate Services: an IRI Perspective, doi:10.1186/2194-6434-1-13.

CHAPTER 5

Erik Tomsen, Jan M Lindsay, Mark Gahegan, Thomas M Wilson, and Daniel M Blake, Evacuation planning in the Auckland Volcanic Field, New Zealand: a spatio-temporal approach for emergency management and transportation network decisions, doi:10.1186/2191-5040-3-6.

CHAPTER 6

Nitesh Bharosa, JinKyu Lee, Marijn Janssen, and H Raghav Rao, An Activity Theory Analysis of Boundary Objects in Cross-Border Information Systems Development for Disaster Management, doi:10.1186/2190-8532-1-15.

CHAPTER 7

Helen Sian Croswel ler, Baneet Arora, Sarah Krystyna Brown, Elizabeth Cottrell,Natalia Irma Deligne, Natalie Ortiz Guerrero, Laura Hobbs, Koji Kiyosugi, Susan Clare Loughlin, Jonathan Lowndes, Martin Nayembil, Lee Siebert, Robert Stephen John Sparks1, Shinji Takarada, and Edward Venzke, Global Database on Large Magnitude Explosive Volcanic Eruptions (Lameve), doi:10.1186/2191-5040-1-4.

CHAPTER 8

Damjan Lešnik, Bojan Lešnik, Jerneja Golub, Miljenko Križmari , **Štefan** Mally and **Štefek** Grmec, Impact of additional module training on the level of basic life support knowledge of first year students at the University of Maribor, doi: 10.1186/1865-1380-4-16.

Index